INFORMATION, UNCERTAINTY, COMPLEXITY

INFORMATION, UNCERTAINTY, COMPLEXITY

J. F. Traub
Departments of Computer Science and Mathematics
Columbia University
New York, New York

G. W. Wasilkowski
Institute of Informatics
University of Warsaw
Warsaw, Poland

H. Woźniakowski
Institute of Informatics
University of Warsaw
Warsaw, Poland
and
Department of Computer Science
Columbia University
New York, New York

1983

ADDISON-WESLEY PUBLISHING COMPANY
Advanced Book Program/World Science Division
Reading, Massachusetts
London · Amsterdam · Don Mills, Ontario · Sydney · Tokyo

Library of Congress Cataloging in Publication Data

Traub, J. F. (Joseph Frederick), 1932-
Information, uncertainty, complexity.

Bibliography: p.
Includes index.
1. Approximation theory--Data processing.
2. Mathematical optimization--Data processing.
3. Computational complexity. I. Wasilkowski, G. W.
II. Woźniakowski, H. III. Title.
QA297.T69 1983 511'.4 82-16428
ISBN 0-201-07890-2

American Mathematical Society (MOS) Subject Classification (1980): 68C25

Published simultaneously in Canada.

Manufactured in the United States of America

ABCDEFGHIJ-HA-898765432

We gratefully dedicate this work
to our parents

Irena Woźniakowska,
Mimi and Leo Traub

and to my wife

Anna Wasilkowska

CONTENTS

PREFACE

We are developing a general framework for studying optimal algorithms and their complexity for any problem which is approximately solved. Since uncertainty in the solution is measured by the parameter ϵ, we refer to this theory as ϵ-complexity. This is the second monograph devoted to the development of this theory. In a previous monograph ("A General Theory of Optimal Algorithms", Academic Press, 1980) uncertainty was measured by a norm. This is an appropriate assumption for "continuous" problems. In this monograph a unified theory, which includes both continuous and discrete problems, is achieved by introducing a more general measure of uncertainty.

Both this and the previous monograph are devoted almost entirely to a worst case setting. We are committed to writing a third volume devoted to the study of various probabilistic settings. Since "asymptotic" methods are widely used, asymptotic models will also be studied in future work.

It gives us great pleasure to acknowledge our debts. C.A. Micchelli and K. Sikorski read the entire manuscript and suggested valuable improvements. We are particularly indebted to A.G. Werschulz who reviewed the manuscript and assisted us with its editing. H.T. Kung made several important suggestions. We are indebted to N.K. Brassfield and M.C. Forry for their admirable typing of a very difficult manuscript under tight time constraints and to K. Calaway for making numerous changes as we revised the manuscript. We are pleased to thank the National Science Foundation for supporting the research reported here. J.F. Traub also wishes to acknowledge the support of the Advanced Research Projects Agency.

J. F. Traub
G. W. Wasilkowski
H. Woźniakowski

INFORMATION, UNCERTAINTY, COMPLEXITY

OVERVIEW

We develop a general mathematical theory on the optimal reduction of uncertainty. We study two basic questions:

1. Can the uncertainty be reduced to a preassigned tolerance?
2. How much must this cost?

For most problems we have to be satisfied with only an approximate solution, that is, we must accept some uncertainty in the solution. Typically, only certain problems with a finite number of inputs and a finite number of outputs, both finitely representable, can be solved exactly. Furthermore, in a distributed environment the solution of even such problems can be only approximated. This motivates our interest in the optimal reduction of uncertainty.

We wish to make two contributions. The first is to supply a general conceptual framework within which people can think about how to deal with uncertainty. Although this framework requires very few basic concepts, very strong results are achieved (see Section 1.2 of Chapter 1). For the first time, very general notions of optimal algorithm and optimal information are formalized.

The second contribution is to create optimal or near-optimal algorithms for numerous diverse application areas. Accomplishing this is a two-stage process. The first stage is to create a number of paradigms which, when applied to a problem, are guaranteed to lead to a good algorithm. The virtue of this approach is that the same paradigm may be successfully applied to many different situations. Examples of such paradigms are the prescriptions for central, interior, and interpolatory classes. The second stage is to use the paradigm to create a good algorithm for a particular problem. This second stage is problem-dependent and can be technically very difficult.

We believe that optimal algorithms and their complexity can be studied as a unified discipline and that the current split between discrete and continuous algorithms is not necessary. In a previous monograph,* uncertainty was measured through a norm. This is appropriate for continuous problems. Here we achieve a synthesis between the discrete and continuous by showing how uncertainty can be measured without a norm or a metric. We work on general sets which can be discrete or continuous. To demonstrate the success of this synthesis, we present many examples drawn from both discrete and continuous mathematics.

We provide a second synthesis. We measure the uncertainty by a nonnegative number ϵ. When the tolerance parameter ϵ is positive, we have ϵ-uncertainty. When it is zero, we have certainty. We do not require a sharp distinction between ϵ positive and zero.

We coin the phrase *ϵ-complexity* to denote the general theory. Earlier, the term *analytic complexity* was used to denote the study of optimality and complexity for continuous problems and positive ϵ. Since the phrase "analytic complexity" carries a connotation of analysis, we shall reserve that designation for continuous problems and use "ϵ-complexity" for the general setting.

There are many measures of ϵ-uncertainty. These include:

1. The worst case: some error criterion is always less than ϵ;
2. The average case: some average-error criterion is less than ϵ;
3. The probabilistic case: some error criterion is less than ϵ probabilistically.

This monograph is devoted almost entirely to the worst case. The average and probabilistic cases will be studied in a future volume.

Problems are solved only approximately for two reasons: either they cannot be solved exactly or a decision is made not to solve them exactly. We discuss both these reasons.

Problems cannot be solved exactly because

1. Information is partial;
2. Information is approximate;
3. The class of algorithms is restricted.

We discuss each of these in turn.

Information is partial when there are problem elements indistinguishable with the available information.

Information can be approximate for many reasons. These include:

a) Computer errors;

---

*J.F. Traub and H. Woźniakowski, *A General Theory of Optimal Algorithms*, Academic Press, New York, 1980. For brevity we shall refer to this monograph as GTOA.

b) Transmission errors;

c) Limitations of number representation and arithmetic;

d) Adversary's lies;

e) Limitations on measurement accuracy. These can be due to instrument limits or they can be intrinsic, as required by Heisenberg uncertainty.

Most real-world problems have access to only partial or approximate information.

Partial and approximate information sets limits on how well even an "idealized" algorithm can reduce uncertainty. An idealized algorithm is *any* mapping on the information. Even with this extremely general idea of algorithm, we can sometimes prove that the ϵ-uncertainty criterion cannot be satisfied. Thus, to obtain negative results we want the most general notion of algorithm. If, on the other hand, we have enough information to meet the ϵ-uncertainty criterion, it is often necessary or desirable to restrict the class of algorithms.

Restricting the class of algorithms can only increase the uncertainty of solving the problem. We introduce two types of restriction: realizable algorithms and permissible algorithms. The difference between these two classes is explained in Chapter 5. Examples of restricted classes of algorithms include:

Turing machine algorithms

Polynomial-time algorithms

Algorithms that are computable functions

On-line algorithms

Algorithms that are linear functions of the input

Stable algorithms

Many of the questions we ask are new. We present a new view of information, algorithms, and complexity and must request the reader's indulgence because new concepts and questions require some new vocabulary. In this Overview we use many technical words without definition. They are defined rigorously and in great generality later. When describing other peoples' results in examples and applications, we often utilize our unifying concepts and terminology rather than the terminology of the author.

Numerous examples are interspersed with theoretical developments to illustrate concepts and results. The last chapter contains major applications. These examples and applications are only a small sample of those that could have been chosen. The examples are:

Binary search

Integration

Normed linear space

Polynomial zeros

Uniform approximation

Division

Most of these examples are continued through the text. The first three examples are the most extensively studied. Binary search and integration were chosen as typical, but relatively simple, examples of discrete and continuous problems. The normed linear space example was chosen to show how the setting of GTOA fits into the new framework.

The applications studied in the last chapter are:

Algebraic coding theory

Distributed computation

Continuous binary search

Bin packing

Nonlinear equations

Linear equations

Database security

Boolean functions

Information theory

Decision theory

Pointwise approximation is studied in Appendix G. Since the normed linear space example is a special case of our setting, all the applications in GTOA provide further instances of applications. These include optimization, approximation, and partial differential equations.

The role of information is central, augmented by the notion of realizable algorithm. Sharp lower bounds on uncertainty are provided by the radii of information. These are three such radii (for exact information, for approximate information, and for the class of realizable algorithms) introduced in the first three chapters, respectively. Because of its relevance, we quote a paragraph from the Overview of GTOA:

> *The generality and power of the theory stems from the central role of information. Adversary arguments based on the information used by an algorithm lead to lower bound theorems. While the notion of informa-*

tion leads to generality it also permits remarkable simplicity. Numerous earlier papers obtain an optimal algorithm under various technical assumptions on the class of algorithms and the class of problem elements. These assumptions are often not verifiable. Our results depend only on the information used and broad verifiable properties of a class of problem elements. They are independent of the structure of an algorithm; they depend only on the information the algorithm uses.

Since information is so central, what is the relation of our theory to information theory? We confine ourselves to a single example of how information theory appears in our setting (see Section 6.9 of Chapter 6). Information theory may play a more important role in some of the probabilistic settings of our future work.

Our theory uses two mathematical models; for convenience we refer to them as models α and β. In model α we study whether ϵ-uncertainty can be attained; in model β we determine how much it must cost. Thus in model α, basic optimality concepts (such as "optimal error algorithm" and "optimal information") are defined independently of the model of computation. Model β consists of model α as well as of concepts that define complexity; the basic optimality concept here is problem complexity. The first four chapters are devoted to model α; only in Chapter 5 do we introduce our model of computation. There are several advantages to this. Negative results can often be proven in model α. For example, we show that ϵ-uncertainty cannot be attained. Since these negative results are independent of the model of computation, they are all the stronger. Furthermore, we must ascertain the optimal error algorithm using optimal information before we can obtain good bounds on the problem complexity.

We make a few remarks concerning structure and format. Chapters 1 and 2 deal with exact and approximate information, while Chapter 3 introduces realizable algorithms that utilize approximate information. Since exact information is a special case of approximate information, the results of Chapter 1 could have been included in Chapter 2. Similarly, the results of Chapter 3 include those of the first two chapters. Although we could have started with Chapter 3, we chose not to, in order to make the material more accessible to the reader. Furthermore, the reader who is interested in exact information only, can read about it without understanding the complications of later material. To clarify relations between corresponding theorems, the first three chapters have similar structure.

So as not to distract the reader from the mainstream of the development, we have moved some of the results to the appendixes; these results are either not essential to the theory or mathematically more difficult.

We describe our system for referring to material within the text. Theorems, equations, remarks, etc. are numbered separately for each section. A reference to material within the same chapter does not name the chapter; a reference to material in a different chapter names the chapter.

This work is self-contained and may be read without reference to GTOA.

The normed linear space setting of GTOA becomes merely an example in this book. However, for the reader who is seriously interested in this setting, GTOA will be of greater relevance. For the reader familiar with GTOA, we present below a partial list of how the earlier work is generalized. We emphasize that the normed linear space setting sometimes gives stronger results than those obtainable in the general setting. Although many of the results of GTOA could be included in the new setting, we limit ourselves to lower and upper bounds, optimal information, and optimal algorithms.

1. In GTOA, norms are used to define concepts such as radius of information and algorithm error. Norms are not used in the new setting.

2. In GTOA, the operators used to define information, algorithm, etc., are defined on spaces, some of which are linear. Here the operators are defined on sets.

3. GTOA is confined to "continuous" problems. Because of items (1) and (2), we can now deal equally well with continuous and discrete problems.

4. In GTOA, the solution element $S(f)$ is an element of a normed linear space, whereas here it is a set. There are many applications for which this is desirable (see, for example, subsection (ii) in Section 6.6 of Chapter 6).

5. In GTOA, the parameter ϵ is defined separately from the solution operator S. Here ϵ is included in the definition of S.

6. Although the case of complete information is formally included in GTOA, the adversary principle is trivial in that case. Only partial information leads to uncertainty. Here information can be complete and still lead to an interesting theory.

7. There is no concept of approximate information in GTOA.

8. There is no concept of realizable algorithm in GTOA.

We briefly summarize the contents of this monograph (see Introduction to each chapter for more detailed summaries).

Chapter 1 The basic concepts are formalized. An adversary principle leads to sharp lower bounds on local and global algorithm error if exact information is available. Algorithms that have minimal local and global error are obtained. The binary search, integration, and normed linear space examples are introduced. These examples will be used throughout to illustrate basic concepts.

Chapter 2 The notion of approximate information is introduced. Sharp lower bounds on algorithm error and optimal algorithms are obtained if only approximate information is available.

Chapter 3 In the first two chapters, an extremely general concept of algorithm was utilized. The notion of realizable algorithm is now introduced. Sharp lower bounds and optimal algorithms are obtained for realizable algorithms that use approximate information.

CHAPTER 4 In the first three chapters, information was regarded as fixed. Information is now varied and optimal information is studied. General conditions are obtained under which nonadaptive information is as powerful or nearly as powerful as adaptive information.

CHAPTER 5 The results of the first four chapters are independent of the model of computation. Our model of computation is defined (it is essentially the same as the one in GTOA) and a variety of examples is used to illustrate the basic complexity concepts.

CHAPTER 6 A variety of application areas is presented to illustrate major concepts. The first three applications (algebraic coding theory, distributed computation, a continuous version of binary search) illustrate optimal information and are independent of complexity notions. The next three applications (bin packing, nonlinear equations, linear equations) also deal with complexity. Additional applications include database security, information theory, and decision theory, the last two for an average case setting.

APPENDIX A In the normed linear space setting, the error of a central algorithm and an interpolatory algorithm can differ at most by a factor of two. In the general setting, the errors can differ by an arbitrary amount. Relations among the errors are explored.

APPENDIX B The concept of interior algorithm is introduced and it is proven that interior algorithms are almost strongly optimal.

APPENDIX C In the normed linear space setting, the diameter of information differs from the radius at most by a factor of two. The concept of diameter of information is generalized and its relation to radius is established under a regularity assumption.

APPENDIX D The concept of nondeterministic algorithm is introduced and tight bounds on the power of nondeterminism are exhibited.

APPENDIX E The normed linear space setting, extensively studied in GTOA, is generalized. A surprising relation between the radii in these two settings is established.

APPENDIX F Throughout this monograph, the uncertainty parameter ϵ is a real number. In this appendix, some of the results are extended to the case where ϵ is an element of a partially ordered set.

APPENDIX G As an application of the partially ordered error criteria presented in the previous appendix, questions concerning optimal information and optimal algorithms for pointwise approximation are posed and solved.

APPENDIX H Additional lower bound results are presented for the case of realizable algorithms.

CHAPTER ONE

EXACT INFORMATION

1.1 INTRODUCTION

The fundamental concepts of solution operator S defined on F, information operator N, and idealized algorithm ϕ are introduced.

We wish to reduce the uncertainty to ϵ, $\epsilon \geq 0$. We avoid introducing a norm (or a metric) by postulating two properties for S.

An *algorithm* ϕ is any operator defined on the set $N(F)$; that is, ϕ has the information N as its input. Such an algorithm is said to be *idealized*. Idealized algorithms are important for two reasons:

1. Even with this very general definition of algorithm we shall prove negative results. For example, we shall obtain results as to when no idealized algorithm exists for attaining ϵ-uncertainty.
2. If one wishes to carry out a computation, then, in general, we must use the "realizable" algorithms of Chapter 3. However, there are many important problems when even an idealized algorithm can be utilized in computation.

In this chapter we assume exact information. In Chapter 2 we shall study approximate information. Information plays a central role in our theory. Knowing only S, F, and N, we can compute a quantity, the radius of information, which is a sharp lower bound on the error of any algorithm for S that uses the information N. Working at the information rather than at the algorithm level provides a vast simplification.

Upper bounds are also studied. We provide a prescription for a *central algorithm*. Central algorithms enjoy a very strong optimal property. We also discuss *interpolatory algorithms*. An interpolatory algorithm is often "simpler" to construct than a central algorithm. In the normed linear space setting, the error of any interpolatory algorithm is within a factor of at most two from being optimal. This property does not hold in the general setting. Properties of interpolatory and interior algorithms, as well as diameter of information, are explored in Appendixes A, B, and C.

Three examples are introduced in this chapter and continued for the remainder of this work. They serve to illustrate basic concepts. Binary search is a relatively simple example which is used to illustrate our results for a typical discrete problem. Integration plays the same role for continuous problems. The third example shows how the normed linear space setting, the subject of GTOA, is a special case of this work.

We summarize the results of this chapter. In Section 1.2, basic concepts, such as solution operator S, information operator N, and ϵ-approximation, are defined. We assume two basic properties of S which will permit us to study ϵ-uncertainty without assuming a normed space. The binary search and integration examples are introduced.

In Section 1.3, we define algorithm and the class of idealized algorithms. The radius of information $r(N)$ is introduced and we establish the basic theorem (Theorem 3.1) that $r(N)$ is a sharp lower bound on the worst-case error of any idealized algorithm that uses the information N. It follows that we cannot reduce the uncertainty to ϵ if the radius of information is greater than ϵ (Corollary 3.1). The *normed linear space setting* is introduced.

In the concluding section we study upper bounds. We introduce central algorithms and prove that an algorithm is strongly optimal iff it is central. Interpolatory and interior algorithms and diameter of information are briefly discussed.

1.2 BASIC DEFINITIONS

Let F and G be two given sets. By 2^G we denote the power set of G, that is, the class of all subsets of G. Consider the operator S such that

$$S : F \times \mathbb{R}_+ \to 2^G, \tag{2.1}$$

where $\mathbb{R}_+ = [0, +\infty)$.

We assume that the operator S has two properties:

$$S(f,0) \neq \emptyset, \quad \forall f \in F, \tag{P$_S$.1}$$

$$\delta_1 \leq \delta_2 \text{ implies } S(f,\delta_1) \subset S(f,\delta_2) \tag{P$_S$.2}$$

for all $\delta_1, \delta_2 \in \mathbb{R}_+$ and all $f \in F$.

Thus the set $S(f,\delta)$ grows as δ increases and $S(f,0)$ is nonempty. These two properties of S will enable us to define the concept of error without introducing a norm in the set G. We call S the *solution operator*. We call $f \in F$ a *problem element*, and any $g \in S(f,0)$ (where $f \in F$) a *solution element*.

Let $\epsilon \geq 0$ be a given number. The parameter ϵ is a measure of the acceptable uncertainty. By an *ϵ-approximation* we mean any element $x \in G$ such that

$$x \in S(f,\epsilon). \tag{2.2}$$

Note that any $x \in G$ satisfying (2.2) will suffice as an ϵ-approximation. Thus $S(f,\epsilon)$ is a subset of G and hence $S(f,\epsilon) \in 2^G$. We illustrate some of these concepts in

Example 2.1 Let

$$F = \{ f\colon f\colon \mathbb{R} \to \mathbb{R},\ f \text{ is continuous and has at least one real zero} \},$$

$$G = \mathbb{R},$$

$$S(f,\epsilon) = \{ x \in G : | f(x) | \leq \epsilon \}.$$

If f has a unique zero, then $S(f,0)$ is a singleton set, while $S(f,\epsilon)$ is not a singleton for $\epsilon > 0$. On the other hand, suppose f has more than one zero. Then for all $\epsilon \geq 0$, $S(f,\epsilon)$ is not a singleton.

Even though this is a simple example, it cannot be represented in the normed linear space setting of GTOA (see Remark 2.2). ■

Remark 2.1 How restrictive are the assumptions concerning S? The first assumption states that there is something to find. With the interpretation that δ measures uncertainty, the second assumption states that as the required uncertainty decreases, the set of elements that satisfies the criterion becomes smaller. Thus we may view these two properties as nonrestrictive. ■

Remark 2.2 We have mentioned that we do not have to introduce a norm in the set G. What kind of structure is induced by properties (P_s.1) and (P_s.2)? Werschulz [82] has shown that, roughly speaking, these properties induce a family of pseudometrics on G. Furthermore, he has shown that a *family* is necessary. That is, there exists an interesting and naturally occurring problem (defined in subsection 6.6(ii) of Chapter 6) that cannot be represented using a *single* pseudometric. ■

Remark 2.3 We choose a real number ϵ as a measure of acceptable uncertainty. The assumption that ϵ is a real number simplifies our analysis and, more importantly, models many practical cases. One may, however, consider the more general case where ϵ is assumed to be an element of a partially ordered set. We show in Appendix F that most of our concepts and results can be generalized to the partially ordered setting. In Appendix G we study the pointwise approximation problem for which it is natural to assume that ϵ is a real function. ■

For any $f \in F$ we want to find an ϵ-approximation. In order to do this we proceed as follows. We assume that, in general, the element f is not known to us. Instead we have some information on f given by $N(f)$, where

$$N : F \to H \tag{2.3}$$

is called an *information operator* (or for brevity *information*) and H is the image set of N. Knowing $N(f)$, we seek an ϵ-approximation. Given S, N, we show how to answer two questions:

1. Is it possible to find an ϵ-approximation?

2. What is the "optimal" algorithm for computing the ϵ-approximation?

In Chapter 2 we consider approximate information.

Remark 2.4 The *total information* is specified by the solution operator S, the sets F and G, and the information operator N. Since in this work S, F, and G are fixed, we refer to N as the information. ■

We illustrate our model by two examples.

Example: Binary Search Let $F = \{1, 2, \ldots, m\}$ and $G = \{1, 2, \ldots\}$. Define

$$S(f,\epsilon) = \{g \in G\colon |f-g| \leq \epsilon m\}$$

as the set of integers whose distance to f does not exceed ϵm. Then (P_s.1) and (P_s.2) hold. Let T_i be a subset of F, where $i = 1,2,\ldots,n$. Let $H = \{0,1\}^n$ and

$$N(f) = [Q(f,T_1),\ Q(f,T_2),\ \ldots,\ Q(f,T_n)]$$

where the question Q is given by

$$Q(f,T_i) = \begin{cases} 1 & \text{if } f \in T_i, \\ 0 & \text{if } f \notin T_i. \end{cases}$$

Thus the problem is to identify an unknown f to within ϵm, knowing whether f belongs to a given set T_i, $i = 1, \ldots, n$. This is called *binary search* because the range of Q consists of two elements. We want to select sets T_i so as to minimize the number of questions required to find an ϵ-approximation. In Chapter 2 we return to this problem assuming that up to k of the questions receive erroneous answers (see also Rivest et al. [80]). This example will be continued in Section 1.4. ■

Example: Integration Let $F = \{f\colon f\colon [0,1] \to \mathbb{R}$, where f is absolutely continuous and $|f'(t)| \leq 1$ a.e.$\}$. Let $G = \mathbb{R}$ and

$$S(f,\epsilon) = \{g \in \mathbb{R} : |g - \textstyle\int_0^1 f(t)\,dt| \leq \epsilon\}$$

be the set of numbers that approximate the integral of f to within ϵ. Then (P_s.1) and (P_s.2) hold.

Let $H = \mathbb{R}^n$ and $N(f) = [f(x_1),\ f(x_2), \ldots,\ f(x_n)]$. Thus we know function values at n points x_i. Knowing $N(f)$, we approximate the integral of f (see GTOA where integration is treated as one of the numerous applications). This example will be continued in Section 1.4. ■

1.3 LOWER BOUNDS

Let

$$V(N,f) = \{\tilde{f} \in F: N(\tilde{f}) = N(f)\} \tag{3.1}$$

be the set of all elements $\tilde{f}$ which have the same information as f. Then, knowing $N(f)$, it is impossible to recognize from which set $S(\tilde{f},\epsilon)$, $\tilde{f} \in V(N,f)$, we have to find an element. It is obvious that we can find an element x that belongs to all sets $S(\tilde{f},\epsilon)$ if and only if their intersection is nonempty. Otherwise our "adversary" can select a set $S(\tilde{f},\epsilon)$ to which x does not belong. This adversary principle is schematized in Fig. 1.1.

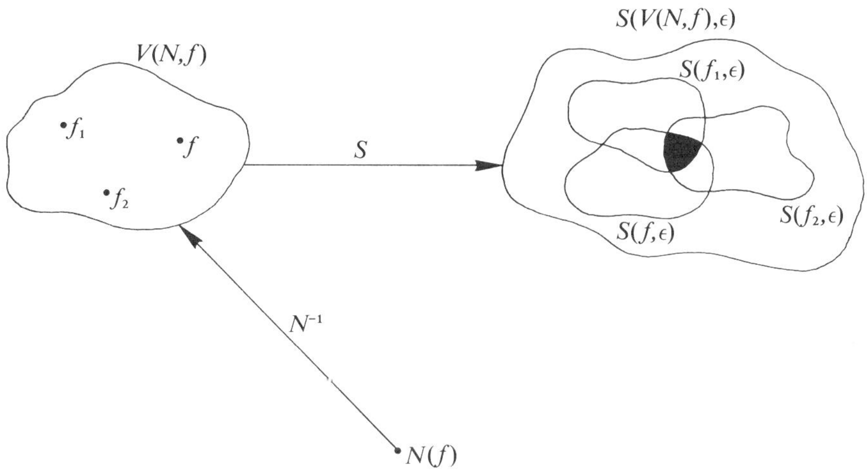

Figure 1.1

The information operator N is called *partial* if N is many-to-one in F and it is called *complete* if it is one-to-one in F. Thus, N is partial iff $V(N,f)$ is not a singleton for some f from F, and N is complete iff $V(N,f)$ is a singleton for every f from F. Note that this adversary principle becomes trivial for complete information since $V(N,f)$ is a singleton and we can find an ϵ-approximation for any $\epsilon \geq 0$.

Let

$$A(N,f,\delta) = \bigcap_{\tilde{f} \in V(N,f)} S(\tilde{f},\delta). \tag{3.2}$$

In Fig. 1.1, $A(N,f,\epsilon)$ is denoted by the darkened area. From (P_s.2) we conclude that $\delta_1 \leq \delta_2$ implies $A(N,f,\delta_1) \subset A(N,f,\delta_2)$.

Let

$$r(N,f) = \inf\{\delta: A(N,f,\delta) \neq \emptyset\}, \tag{3.3}$$

assuming by convention that $\inf \emptyset = +\infty$. Note that $r(N,f)$ depends on S and might be written as $r(N,S,f)$. Since S is fixed, we choose to simply write $r(N,f)$. We call $r(N,f)$ the *local radius*. Thus $r(N,f)$ is roughly the smallest δ for which there exists an element belonging to $S(\tilde{f},\delta)$ for all $\tilde{f}$ that have the same information as f. We define the (global) *radius of information* as

$$r(N) = \sup_{f \in F} r(N,f) \qquad (= \inf\{\delta: A(N,f,\delta) \neq \emptyset, \forall f \in F\}). \tag{3.4}$$

Note that $r(N)$ depends on S and might be written $r(N,S)$. Since S is fixed, we need write only $r(N)$. We illlustrate this concept by an example which justifies calling $r(N,f)$ and $r(N)$ radii.

Example: Normed Linear Space We show how the normed linear space setting studied in GTOA fits into the setting of this work (see also Appendix E where we deal with a generalization of this setting).

Let F be a subset of a linear space over the real or complex field. Let G be a normed linear space over the real or complex field. Let $S: F \to G$ be an operator. Note that the domain and range of the operator S just defined differ from the operator denoted by S up to this point. This is done for consistency with GTOA. We believe this abuse of notation will not lead to any confusion.

Define

$$S(f,\epsilon) = \begin{cases} \{g \in G: \| S(f) - g \| < \epsilon\} & \text{for} \quad \epsilon > 0, \\ \{S(f)\} & \text{for} \quad \epsilon = 0. \end{cases}$$

Thus $S(f,\epsilon)$ is the ball with center $S(f)$ and radius ϵ. Note that $S(f,0)$ is a singleton. It is obvious that $(P_S.1)$ and $(P_S.2)$ hold. An ϵ-approximation with $\epsilon > 0$ is now an element x such that $\|S(f) - x\| < \epsilon$.

In this case

$$A(N,f,\delta) = \begin{cases} \bigcap_{\tilde{f} \in V(N,f)} \{g: \| S(\tilde{f}) - g \| < \delta\}, & \delta > 0, \\ \bigcap_{\tilde{f} \in V(N,f)} \{S(\tilde{f})\} & , \quad \delta = 0. \end{cases}$$

Then

$$r(N,f) = \inf_{g \in G} \sup_{\tilde{f} \in V(N,f)} \| S(\tilde{f}) - g \|$$

is essentially the minimal radius of a ball containing $S(V(N,f))$. Similarly, $r(N)$ is the radius of $S(V(N,f))$ for a worst f. This example will be continued in Section 4. ■

We present a second example to illustrate the radius of information.

Example 3.1 Suppose we want to find the minimal amount of information needed for an exact solution. This corresponds to $\epsilon = 0$. For instance, consider the minimal number of questions to exactly identify the element f in the Binary Search Problem (see, for example, Abelson [80], Rabin [72], Rivest and Vuillemin [76], and Yao [79] as well as Chapter 6). We now show how this fits into our model.

For given sets F and G, let $S : F \to G$. Define

$$S(f,\epsilon) = \{S(f)\}, \quad \forall \epsilon \geq 0.$$

Thus, $S(f,\epsilon)$ is independent of ϵ and $x = S(f)$ is a unique ϵ-approximation. Trivially, $(P_s.1)$ and $(P_s.2)$ hold. Let N be given. Then $A(N,f,\delta) = \{S(f)\}$, $\forall \delta \geq 0$, iff $S(\tilde{f}) = S(f)$, $\forall \tilde{f} \in V(N,f)$, that is, $S(V(N,f))$ is a singleton. Otherwise $A(N,f,\delta) = \emptyset$, $\forall \delta \geq 0$. Hence

$$r(N,f) = \begin{cases} 0 & \text{if } S(V(N,f)) \text{ is a singleton,} \\ +\infty & \text{otherwise,} \end{cases}$$

and consequently

$$r(N) - \begin{cases} 0 & \text{if } S(V(N,f)) \text{ is a singleton, } \forall f \in F, \\ +\infty & \text{otherwise.} \end{cases}$$

In this general setting we seek the "minimal amount" of information which makes $r(N)$ zero. ■

To solve (2.2) we use an *algorithm* which is an operator ϕ such that

$$\phi : N(F) \to G. \tag{3.5}$$

Let $\Phi(N)$ be the class of *all* algorithms that use the information operator N. We shall refer to $\Phi(N)$ as the class of *idealized algorithms*.

As we shall see, even with this extremely general idea of algorithm we shall prove negative results. This makes the negative results even stronger. If one wishes to carry out a computation, then in general the class of algorithms must be restricted to a class of "realizable" algorithms. However, there are many important examples where algorithms that are "optimal" in the idealized class are also optimal in a realizable class (see Chapters 3 and 6).

Remark 3.1 As in Remark 2.4, we stress that an algorithm ϕ uses the *total information* which is specified by the solution operator S, the sets F and G, and the information operator N. Thus $\phi = \phi(N(f); S, F, G, N)$. Since $S, F, G,$ and N are fixed, we need write only $\phi = \phi(N(f))$. ■

Remark 3.2 Here we assume that $\phi(N(f))$ is an element of G. In Appendix D we generalize this by assuming that $\phi(N(f))$ is a subset of G consisting of k elements. If $k > 1$, then ϕ is called a nondeterministic algorithm. We learn in Appendix D whether nondeterminism helps in finding an ϵ-approximation. ■

We now define local and global algorithm errors. Note that G need not be equipped with a norm. For any algorithm ϕ, where $\phi \in \Phi(N)$, let

$$e(\phi,N,f) = \inf\{\delta: \phi(N(f)) \in A(N,f,\delta)\} \tag{3.6}$$

be the *local error of* ϕ and let

$$e(\phi,N) = \sup_{f\in F} e(\phi,N,f) \qquad (=\inf\{\delta: \phi(N(f)) \in S(f,\delta), \forall f \in F\}) \tag{3.7}$$

be the (global) *error of* ϕ. Thus, the local error is roughly the smallest δ for which $\phi(N(f))$ is an element belonging to $S(\tilde{f},\delta)$ for all $\tilde{f}$ that have the same information as f. This is shown in Fig. 1.2 with $\delta > e(\phi,N,f)$, where the darkened region denotes $A(N,f,\delta)$.

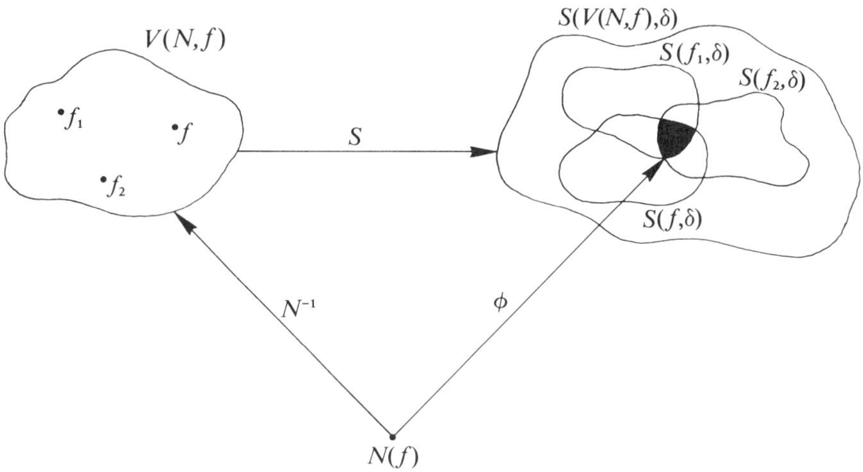

Figure 1.2

It is obvious that the local radius and the radius of information are the sharp lower bounds on the local and global errors of any algorithm. A formal proof is provided by

THEOREM 3.1

$$\inf_{\phi\in\Phi(N)} e(\phi,N,f) = r(N,f), \quad \forall f \in F, \tag{3.8}$$

$$\inf_{\phi\in\Phi(N)} e(\phi,N) = r(N). \tag{3.9}$$

■

PROOF Let $f \in F$ and $\phi \in \Phi(N)$. Without loss of generality assume that $a = e(\phi,N,f) < +\infty$. From (3.6) we conclude that $A(N,f,a + \delta) \neq \emptyset$, $\forall\delta > 0$. Then (3.3) yields $r(N,f) \leq a + \delta$; since δ is arbitrary, $r(N,f) \leq e(\phi,N,f)$. Since ϕ is arbitrary, we have

$$r(N,f) \leq \inf_{\phi\in\Phi(N)} e(\phi,N,f). \tag{3.10}$$

To prove equality, take any positive δ and observe that $A(N,f,r(N,f) + \delta) \neq \emptyset$. Define the algorithm ϕ_δ in such a way that $\phi_\delta(N(f)) \in A(N,f,r(N,f) + \delta)$.
Then

$$e(\phi_\delta,N,f) \leq r(N,f) + \delta\,. \tag{3.11}$$

Since δ is arbitrary, this proves that $\inf_\phi e(\phi,N,f) \leq r(N,f)$. Due to (3.10), this completes the proof of (3.8). Since (3.9) easily follows from (3.8) and (3.11), Theorem 3.1 is proven. ■

Theorem 3.1 enables us to verify when we can solve our problem, i.e., when we can find an ϵ-approximation. It is easy to establish

COROLLARY 3.1 It is possible to find an ϵ-approximation for all $f \in F$ using an information operator N iff

(i) $r(N) < \epsilon$, or

(ii) $r(N) = \epsilon$ and the infimum of (3.9) is attained for an algorithm ϕ and for this algorithm the infimum of (3.7) is also attained.

■

PROOF Suppose we can find an ϵ-approximation for all $f \in F$. This means there exists an algorithm ϕ such that $\phi(N(f)) \in S(f,\epsilon)$, $\forall f \in F$. Thus $e(\phi,N) \leq \epsilon$. Since $r(N) \leq e(\phi,N)$ due to (3.9), condition (i) or (ii) holds.

Suppose now that (i) holds. Then there exists an algorithm ϕ such that $e(\phi,N) < \epsilon$ and $\phi(N(f))$ is an ϵ-approximation for all $f \in F$. If (ii) holds, take the algorithm ϕ for which the infimum in (3.9) is attained. Then $e(\phi,N) = r(N) = \epsilon$ and $\phi(N(f))$ is an ϵ-approximation for all $f \in F$. This completes the proof. ■

Remark 3.3 Recall that ϵ measures the acceptable uncertainty. The radius $r(N)$ is a quantitative measure of uncertainty. It is determined by the strength of the total information.

If $r(N) > \epsilon$, the uncertainty cannot be reduced to ϵ; it has to be at least $r(N)$. ■

Remark 3.4 Let N be the zero operator, $N = 0$. Then $r(0)$ is the uncertainty if we know only the solution operator S and the sets F and G. In many cases, $r(0) = \infty$ (see GTOA). By comparing $r(0)$ with $r(N)$, we can see the amount by which N reduces uncertainty. ■

In Section 1.4 we calculate the local radii and the radii of information for the two problems presented in Section 1.2.

1.4 UPPER BOUNDS

In Section 1.3 we proved that the local radius and the radius of information are sharp lower bounds on the local and global errors of any algorithm. In this section we derive algorithms that have minimal or almost minimal local and global errors.

An algorithm ϕ, $\phi \in \Phi(N)$, is called a *strongly optimal error algorithm* iff

$$e(\phi,N,f) = r(N,f), \qquad \forall f \in F, \tag{4.1}$$

and it is called an *optimal error algorithm* iff

$$e(\phi,N) = r(N). \tag{4.2}$$

A strongly optimal error algorithm is also an optimal error algorithm but the converse is, in general, not true. Of course, optimal and strongly optimal is with respect to a particular N. To find a strongly optimal error algorithm observe that, due to (3.3), $A(N,f,\ r(N,f) + \delta)$ is nonempty for all positive δ. Suppose there exists an element $c(f)$ such that

$$c(f) \in A(N,f,\ r(N,f) + \delta), \qquad \forall \delta > 0. \tag{4.3}$$

This element $c(f)$ is called a *center* of f and the algorithm

$$\phi^c(N(f)) = c(f) \tag{4.4}$$

is called a *central algorithm*.

THEOREM 4.1 An algorithm ϕ is a strongly optimal error algorithm iff ϕ is central. ∎

PROOF Let ϕ be strongly optimal. Then (3.6) and (4.1) yield

$$\phi(N(f)) \in A(N,f,\ r(N,f) + \delta), \qquad \forall \delta > 0.$$

Due to (4.3), this means that $\phi(N(f))$ is a center of f and thus ϕ is a central algorithm.

Now let ϕ be central. Then (4.3) and (3.6) yield that $e(\phi,N,f) \leq r(N,f)$. From (3.8) of Theorem 3.1 we conclude that $e(\phi,N,f) = r(N,f)$. Thus ϕ is a strongly optimal error algorithm. This completes the proof. ∎

We illustrate the concept of central algorithm by an example.

Example: Normed Linear Space (continued) An element $c(f)$ now satisfies

$$\sup_{\tilde{f} \in V(N,f)} \| S(\tilde{f}) - c(f) \| = \inf_{x \in G} \sup_{\tilde{f} \in V(N,f)} \| S(\tilde{f}) - x \|,$$

which means that, in the language of the theory of approximation, $c(f)$ is a center of the set $S(V(N,f))$. Thus the concept of central algorithm introduced here coincides with the one introduced in GTOA (p.14) for the special case of the normed linear space setting. Note also that a center of set A need not exist, be unique, or lie in A. This example will be continued in Chapter 3. ∎

Central algorithms enjoy the very strong property of minimizing the local error for *every* problem f. Thus even if the radius of information $r(N)$ is large, it can happen, due to some favorable properties of f, that the radius $r(N,f)$ is much smaller than $r(N)$ and a central algorithm ϕ^c recognizes this favorable situation and produces an element with local error $e(\phi^c,N,f) = r(N,f) << r(N)$.

Further relations between central algorithms and the radius of information are presented in Appendix H. We illustrate the concepts of central algorithm and radius of information by two examples.

Example: Binary Search (continued) Assume that $m = pn$ for an integer p. For simplicity we define sets T_i which serve only to illustrate our concepts. The selection of sets T_i that minimize the number of questions required to find an ϵ-approximation is discussed in Chapters 4 and 6. Let

$$T_i = \{(i-1)p + 1,\ldots,(i-1)p + p\}, \qquad i = 1,2,\ldots,n.$$

Then $N(f) = [0,\ldots,0,\underset{j}{1},0,\ldots,0]$, where $j = j(f) = \lceil f/p \rceil$. From (3.1) and (3.2) we have $V(N,f) = T_j$ and $A(N,f,\delta) = \{f_2 - a, f_2 - a + 1,\ldots, f_1 + a\} \cap G$, where $a = \lfloor \delta m \rfloor$ and $f_1 = (j-1)p+1$, $f_2 = (j-1)p+p$. Thus

$$r(N) = r(N,f) = \begin{cases} \dfrac{1}{2n} & \text{if} \quad p \text{ is even,} \\[2ex] \dfrac{1}{2n} - \dfrac{1}{2m} & \text{if} \quad p \text{ is odd.} \end{cases}$$

Define two algorithms

$$\phi_k(N(f)) = \begin{cases} (j-1)p + \dfrac{p}{2} + (k-1) & \text{if } p \text{ is even,} \\[2ex] (j-1)p + \dfrac{p+1}{2} & \text{if } p \text{ is odd} \end{cases}$$

for $k = 1,2$. Then

$$e(\phi_k, N, f) = r(N,f),$$

which shows that the algorithms ϕ_k are central. We can find an ϵ-approximation iff

$$\epsilon \geq \begin{cases} \dfrac{1}{2n} & \text{if } p \text{ is even,} \\[2ex] \dfrac{1}{2n} - \dfrac{1}{2m} & \text{if } p \text{ is odd.} \end{cases}$$

This example will be continued in Chapter 2. ■

Example: Integration (continued) Assume for simplicity that $n \geq 2$ and that points x_i are equally spaced, $x_i = (i-1)/(n-1)$, $i = 1,2,\ldots,n$. Define

$$w_i = \frac{2i-1}{2(n-1)} - \frac{f(x_{i+1}) - f(x_i)}{2}, \qquad z_i = \frac{2i-1}{2(n-1)} + \frac{f(x_{i+1}) - f(x_i)}{2}$$

for $i = 1,2,\ldots,n$, and two functions

$$f^-(x) = \begin{cases} f(x_i) - x + x_i & \text{if} \quad x_i \leq x \leq w_i, \\[1ex] f(x_{i+1}) + x - x_{i+1} & \text{if} \quad w_i \leq x \leq x_{i+1}, \end{cases}$$

$$f^+(x) = \begin{cases} f(x_i) + x - x_i & \text{if} \quad x_i \leq x \leq z_i, \\[1ex] f(x_{i+1}) - x + x_{i+1} & \text{if} \quad z_i \leq x \leq x_{i+1}. \end{cases}$$

Then f^- and f^+ belong to $V(N,f)$ and

$$f^-(x) \le \tilde{f}(x) \le f^+(x), \qquad \forall x \in [0,1], \qquad \forall \tilde{f} \in V(N,f).$$

From this we get

$$a = \int_0^1 f^-(x)\,dx \le \int_0^1 \tilde{f}(x)\,dx \le b = \int_0^1 f^+(x)\,dx.$$

Thus

$$A(N,f,\delta) = \left[\frac{b+a}{2} - \left(\delta - \frac{b-a}{2}\right), \frac{b+a}{2} + \delta - \frac{b-a}{2}\right]$$
$$= [b - \delta,\ a + \delta]$$

(with the convention that $[c,d] = \emptyset$ if $c > d$).

This implies that

$$r(N,f) = \tfrac{1}{2}(b - a) = \tfrac{1}{2}\int_0^1 [f^+(x) - f^-(x)]\,dx.$$

It is maximized if $f(x_i) = 0$, that is,

$$r(N) = r(N,f) = \frac{1}{4(n-1)},$$

where $f \in V(N,0)$. Defining

$$\phi(N(f)) = \tfrac{1}{2}(b + a) = \tfrac{1}{2}\int_0^1 [f^+(x) + f^-(x)]\,dx,$$

we get $e(\phi,N,f) = r(N,f)$. Thus ϕ is a unique central algorithm. This example will be continued in Chapter 2. ∎

For the normed linear space setting we can define a second class of algorithms which are almost strongly optimal. This is the class of interpolatory algorithms defined as follows. Choose an element $\tilde{f}$ indistinguishable from f under N. That is, choose $\tilde{f} \in F$ such that $N(\tilde{f}) = N(f)$ and such that $S(\tilde{f})$ is "simple" to calculate. Then we define $\phi^I(N(f)) = S(\tilde{f})$. In the normed linear space setting, interpolatory algorithms are important in many applications (see GTOA).

The most important property of $\phi^I(N(f))$ in a normed linear space is that

$$e(\phi^I,N,f) \le \text{diam}(N,f) \le 2r(N,f), \qquad \forall f \in F, \tag{4.5}$$

where diam(N,f), called the *local diameter of information,* is equal to

$$\text{diam}(N,f) = \sup_{f_1,f_2 \in V(N,f)} \| S(f_1) - S(f_2) \|. \tag{4.6}$$

Therefore ϕ^I is almost strongly optimal. On the other hand, it is often far easier to compute than ϕ^c.

Can the definition of interpolatory algorithm be extended to the setting of this work and does (4.5) still hold? We shall give a natural generalization of interpolatory algorithm but shall show that (4.5) need not hold. Indeed the ratio of the local error of an interpolatory algorithm to the local radius may be unbounded. Interpolatory algorithms enjoy a property close to (4.5) for certain S, which we characterize in Appendix A. Interpolatory algorithms are of interest for these S.

Since interpolatory algorithms are not always strongly optimal, we introduce in Appendix B a different class of algorithms called *interior algorithms*. We show that they are always almost strongly optimal. Relations between central, interpolatory, and interior algorithms are also exhibited in Appendix B.

The local diameter $d(N,f)$ of information, which is defined by (4.6) for the normed linear space setting, is conceptually much simpler to compute than the radius of information. Furthermore, in many cases it is enough to know $d(N,f)$ since

$$r(N,f) \le d(N,f) \le 2r(N,f). \tag{4.7}$$

Can the definition of diameter of information be extended to the setting of this work and does (4.7) hold? We address this problem in Appendix C. We show that the right-hand inequality of (4.7) always holds, while the left-hand inequality of (4.7) holds under a "regularity" assumption on S.

We have placed the study of interpolatory and interior algorithms as well as of diameter of information in Appendixes A, B and C since it is technically rather difficult to establish these results and since we do not wish to distract the reader from the mainstream of this work.

CHAPTER TWO

APPROXIMATE INFORMATION

2.1 INTRODUCTION

We assume that only approximate information N_ρ is available, where the parameter ρ is a measure of error. When $\rho = 0$, all definitions and results reduce to the exact information setting of Chapter 1.

There are numerous instances of approximate information. They include:

1. Adversary situations where incorrect information is intentionally presented.
2. Unintentional information errors due, for example, to computer or communication errors.
3. Information errors inherent in representation limitations of numbers. An example is provided by finite-precision representation and arithmetic.

Many of our examples and applications involve approximate information. They include:

1. Binary search with up to k erroneous answers (Section 2.2).
2. Integration with perturbed function values (Section 2.2).
3. Transmission of a message across a noisy channel (Section 6.2 of Chapter 6).
4. Continuous binary search with up to k erroneous answers (Section 6.4 of Chapter 6).
5. Linear systems with matrix represented in finite-precision arithmetic (subsections (ii) and (iii) of Section 6.7, Chapter 6.)

Examples of approximate information were considered by Marchuk and Osipenko [75], Micchelli [75], and Micchelli and Rivlin [77].

We summarize the results of this chapter. In Section 2.2 the basic concepts of approximate information operator N_ρ and information error operator E are defined. We assume two basic properties of E which relieve us of assuming a normed space. The binary search example is considered with up to k erroneous answers, while the integration example is considered with perturbed function values. In Sections 2.3 and 2.4 we extend the results of the corresponding sections of Chapter 1 to approximate information.

2.2 APPROXIMATE INFORMATION

Let N, $N : F \to H$, be an information operator. In this chapter we assume that $N(f)$ is not available. Instead we are given an element $N_\rho(f)$ which represents approximate information on f. We formalize this as follows. Consider an operator E such that

$$E : H \times \mathbb{R}_+ \to 2^H. \tag{2.1}$$

We assume that the operator E has two properties:

$$E(h,0) = \{h\}, \qquad \forall h \in H, \tag{P$_E$.1}$$

$$\delta_1 \le \delta_2 \text{ implies } E(h,\delta_1) \subset E(h,\delta_2) \tag{P$_E$.2}$$

for all $\delta_1, \delta_2 \in \mathbb{R}_+$ and all $h \in H$.

The operator E has properties similar to the solution operator S. The only difference is that $E(h,0)$ is a singleton, whereas $S(f,0)$ is nonempty. The set $E(N(f),\delta)$ represents approximate information with error which is measured by a nonnegative parameter δ. For $\delta = 0$, $E(N(f),0) = \{N(f)\}$ becomes the exact information. We call E the *information error operator.* Consider an operator $N_\rho : F \to H$ such that

$$N_\rho(f) \in E(N(f),\rho), \qquad \forall f \in F, \tag{2.2}$$

for some fixed $\rho \ge 0$. We assume that $N_\rho(f)$ is given. This represents approximate information on f, and ρ is a measure of error. Note that $N_0(f) = N(f)$; this case was studied in Chapter 1. We call N_ρ the *approximate information operator.* We illustrate our model by two examples.

Example: Binary Search (continued) Recall that

$$N(f) = [Q(f, T_1),\dots, Q(f, T_n)],$$

where $Q(f, T_i)$ is the answer to the question whether f belongs to the set T_i.

Suppose that up to k of the answers may be erroneous, where $k \leq n$. This means that the vector

$$N_k(f) = [y_1, y_2, \ldots, y_n] \in H = \{0,1\}^n$$

differs from $N(f)$ in at most k places. For two vectors $x,y \in H$ denote

$$\mathrm{HAM}(x,y) = \mathrm{card}\{i: x_i \neq y_i\}$$

as the number of components in which x and y differ. Here $\mathrm{HAM}(x,y)$ is the Hamming distance between vectors x, y. Define

$$E(h,\delta) = \{g \in H: \mathrm{HAM}(h,g) \leq \delta\} .$$

Clearly, ($\mathrm{P_E}$.1) and ($\mathrm{P_E}$.2) hold. Then $N_k(f) \in E(N(f),k)$, which means that (2.2) holds with $\rho = k$ (see Rivest et al. [80]). This example will be continued in Section 3. ■

Example: Integration (continued) Recall that

$$N(f) = [f(x_1), f(x_2), \ldots, f(x_n)].$$

Suppose that instead of the exact function values $f(x_i)$ we are given $y_i = y_i(f)$ such that

$$| f(x_i) - y_i | \leq \rho(|y_i| + \eta) \tag{2.3}$$

for some nonnegative ρ and η. For $|y_i| >> \eta$, this means that the relative error of y_i and $f(x_i)$ is roughly ρ, whereas for $|y_i| << \eta$ the absolute error of y_i and $f(x_i)$ is roughly $\rho\eta$. Thus (2.3) is the combined relative and absolute error of y_i and $f(x_i)$. Define

$$E(h,\delta) = \{g \in \mathbb{R}^n : |h_i - g_i| \leq \delta(|g_i| + \eta)\} .$$

Clearly ($\mathrm{P_E}$.1) and ($\mathrm{P_E}$.2) hold and

$$N_\rho(f) = [y_1(f), \ldots, y_n(f)] \in E(N(f),\rho) .$$

This example will be continued in Section 2.3. ■

2.3 LOWER BOUNDS

Let N_ρ be an approximate information operator. Define

$$V(N_\rho,f) = \{\tilde{f} \in F : N_\rho(f) \in E(N(\tilde{f}),\rho)\} \tag{3.1}$$

as the set of all elements $\tilde{f}$ for which $N_\rho(f)$ can be given as approximate information on $\tilde{f}$. Due to (2.2), $f \in V(N_\rho,f)$, and from (3.1) of Chapter 1 we have

$$V(N,f) = V(N_0,f) \subset V(N_\rho,f), \qquad \forall f \in F. \tag{3.2}$$

Knowing $N_\rho(f)$ it is impossible to recognize from which set $S(\tilde{f},\epsilon)$, $\tilde{f} \in V(N_\rho,f)$, we have to find an element. As in Section 1.3 of Chapter 1 we can find an ϵ-approximation iff the intersection of $S(\tilde{f},\epsilon)$ is nonempty. Let

$$A(N_\rho,f,\delta) = \bigcap_{\tilde{f} \in V(N_\rho,f)} S(\tilde{f},\delta) \tag{3.3}$$

and let

$$r(N_\rho,f) = \inf\{\delta : A(N_\rho,f,\delta) \neq \emptyset\} \tag{3.4}$$

be the *local radius of approximate information.* From (3.2) of Chapter 1 and (3.2) it follows that

$$\begin{aligned} A(N_\rho,f,\delta) &\subset A(N_0,f,\delta) = A(N,f,\delta), \\ r(N_\rho,f) &\geq r(N_0,f) = r(N,f), \qquad \forall f \in F. \end{aligned} \tag{3.5}$$

We define the (global) *radius of approximate information* as

$$r(N_\rho) = \sup_{f \in F} r(N_\rho,f) \qquad (= \inf\{\delta: A(N_\rho,f,\delta) \neq \emptyset, \ \forall f \in F\}). \tag{3.6}$$

Clearly, (3.4) of Chapter 1 and (3.5) yield

$$r(N_\rho) \geq r(N_0) = r(N). \tag{3.7}$$

The approximate information operator N_ρ is called *partial* if $V(N_\rho,f)$ is not a singleton for some f from F, and it is called *complete* if $V(N_\rho,f)$ is a singleton for every f from F. Note that if the exact information operator N is partial, then N_ρ is also partial since $V(N,f) \subset V(N_\rho,f)$. If N is complete, then it can happen that N_ρ is still partial or, as we shall see in Section 6.2 of Chapter 6, it can happen that N_ρ is also complete. Of course, if N_ρ is complete, then $r(N_\rho) = 0$.

We shall show that $r(N_\rho,f)$ and $r(N_\rho)$ are sharp lower bounds on the local and global errors of any algorithm that uses the approximate information N_ρ. By an *algorithm* we now mean an operator ϕ such that

$$\phi : N_\rho(F) \to G. \tag{3.8}$$

Let $\Phi(N_\rho)$ be the class of *all* algorithms that use N_ρ. We shall refer to $\Phi(N_\rho)$

as the class of *idealized algorithms.* "Realizable" algorithms will be studied in Chapters 3 and 6. As in Chapter 1, we define

$$e(\phi,N_\rho,f) = \inf\{\delta:\ \phi(N_\rho(f)) \in A(N_\rho,f,\delta)\} \tag{3.9}$$

as the *local error* and

$$e(\phi,N_\rho) = \sup_{f\in F} e(\phi,N_\rho,f) \qquad (= \inf\{\delta : \phi(N_\rho(f)) \in A(N_\rho,f,\delta),\ \forall f \in F\}) \tag{3.10}$$

as the (global) *error*.

THEOREM 3.1

$$\inf_{\phi\in\Phi(N_\rho)} e(\phi,N_\rho,f) = r(N_\rho,f), \qquad \forall f \in F, \tag{3.11}$$

$$\inf_{\phi\in\Phi(N_\rho)} e(\phi,N_\rho) = r(N_\rho)\ . \tag{3.12}$$

■

We omit the proof since it is similar to the proof of Theorem 3.1 of Chapter 1. From Theorem 3.1 we immediately obtain

COROLLARY 3.1 It is possible to find an ϵ-approximation for all $f \in F$ using an approximate information operator N_ρ iff

(i) $r(N_\rho) < \epsilon$, or

(ii) $r(N_\rho) = \epsilon$ and the infimum of (3.12) is attained for an algorithm ϕ and for this algorithm the infimum of (3.10) is also attained.

■

The set $E(N(f),\rho)$ consists of elements representing approximate information on f, and each element of this set can be taken as $N_\rho(f)$. In practice we cannot control which element from $E(N(f),\rho)$ is given as $N_\rho(f)$. Therefore it is interesting to investigate the radius of approximate information in terms of $E(N(f),\rho)$ instead of $N_\rho(f)$. Let b be the *local radius of worst-case approximate information,* i.e.,

$$b = \sup_{N_\rho\,:\,N_\rho(f)\in E(N(f),\rho)} r(N_\rho,f)\ .$$

Then b depends on $N,S,$ and $E(N(f),\rho)$. To stress this dependence, we write

$$r(N_{\rho,E},f) = b = \sup_{N_\rho\,:\,N_\rho(f)\in E(N(f),\rho)} r(N_\rho,f)\ . \tag{3.13}$$

Remark 3.1 If the supremum in (3.13) is attained, then $N_{\rho,E}$ denotes a "worst-case approximate information operator." If the supremum is not attained, there is no such operator, but we shall use the notation $N_{\rho,E}$ as a suggestive device. ■

It is obvious that

$$r(N_{0,E},f) = r(N_0,f) = r(N,f), \tag{3.14}$$

$$\rho_1 \le \rho_2 \text{ implies } r(N_{\rho_1,E},f) \le r(N_{\rho_2,E},f), \qquad \forall f \in F. \tag{3.15}$$

We call

$$r(N_{\rho,E}) = \sup_{f \in F} r(N_{\rho,E},f) \qquad (= \sup_{N_\rho} r(N_\rho)) \tag{3.16}$$

the (global) *radius of worst-case approximate information.* From (3.14) and (3.15) we have

$$r(N_{0,E}) = r(N_0) = r(N) , \tag{3.17}$$

$$\rho_1 \le \rho_2 \text{ implies } r(N_{\rho_1,E}) \le r(N_{\rho_2,E}) . \tag{3.18}$$

This means that $r(N_{\rho,E})$ is a nondecreasing function of ρ. For fixed ρ we have

$$r(N) \le r(N_\rho) \le r(N_{\rho,E}) \tag{3.19}$$

for any approximate information operator N_ρ. As we shall see, both bounds in (3.19) are sharp, i.e., it may happen that $r(N_\rho) = r(N)$ for an approximate information operator N_ρ and $r(N_\rho) = r(N_{\rho,E})$ for another approximate information operator N_ρ. It may also happen that $r(N_{\rho,E})$ is significantly larger than $r(N_\rho)$ for some approximate information N_ρ.

We shall now show that $r(N_{\rho,E},f)$ and $r(N_{\rho,E})$ are the sharp lower bounds on the local and global errors of any algorithm that uses any $N_\rho(f)$. More precisely, let

$$\phi : E(N(F),\rho) \to G \tag{3.20}$$

be an algorithm. For any operator N_ρ such that $N_\rho(f)$ belongs to $E(N(f),\rho)$ we wish $\phi(N_\rho(f))$ to be an ϵ-approximation for all elements $\tilde{f}$ for which $N_\rho(f) \in E(N(\tilde{f}),\rho)$. We define the *local error* as

$$e(\phi,N_{\rho,E},f) = \sup_{N_\rho \,:\, N_\rho(f) \in E(N(f),\rho)} e(\phi N_\rho,f)$$

$$(= \sup_{N_\rho \,:\, N_\rho(f) \in E(N(f),\rho)} \inf\{\delta : \phi(N_\rho(f)) \in A(N_\rho,f,\delta)\}) . \tag{3.21}$$

The (global) *error* is now defined as

$$e(\phi,N_{\rho,E}) = \sup_{f\in F} e(\phi,N_{\rho,E},f) \qquad (= \sup_{N_\rho} e(\phi,N_\rho)) \ . \tag{3.22}$$

Let $\Phi(N_{\rho,E})$ be the class of idealized algorithms, i.e., the class of *all* algorithms ϕ defined by (3.20).

THEOREM 3.2

$$\inf_{\phi\in\Phi(N_{\rho,E})} e(\phi,N_{\rho,E},f) = r(N_{\rho,E},f), \qquad \forall f\in F, \tag{3.23}$$

$$\inf_{\phi\in\Phi(N_{\rho,E})} e(\phi,N_{\rho,E}) = r(N_{\rho,E}). \tag{3.24}$$

■

The proof is straightforward and similar to the proof of Theorem 3.1 of Chapter 1 and therefore is omitted. From Theorem 3.2 we easily obtain

COROLLARY 3.2 It is possible to find an ϵ-approximation for all $f \in F$ using *any* approximate information N_ρ iff

(i) $r(N_{\rho,E}) < \epsilon$, or

(ii) $r(N_{\rho,E}) = \epsilon$ and the supremum in (3.16) or in (3.13) is *not* attained, or

(iii) $r(N_{\rho,E}) = \epsilon$ and the suprema in (3.16) and (3.13) *are* attained for N_ρ for which (ii) of Corollary 3.1 holds. ■

We end this section by calculating the corresponding radii for two examples.

Example: Binary Search (continued) We use definitions and notation from Chapter 1. Assume for simplicity that $k = \rho = 1$ and p is even. Consider first an approximate information operator N_1^1 defined as

$$N_1^1(f) = [0,\ldots,\underset{j}{1},\ldots,0] \ , \qquad j = j(f), \qquad \forall f \in F.$$

Any vector whose Hamming distance from $N_1^1(f)$ is unity cannot be correct information $N(\tilde{f})$ since $N(\tilde{f})$ must have a single 1. Therefore we conclude that $N_1^1(f)$ is correct. Then $V(N_1^1, f) = V(N,f)$, which yields

$$r(N_1^1) = r(N_1^1,f) = r(N) = \frac{1}{2n} \ .$$

Suppose now that an approximate information operator N_1^2 is given by

$$N_1^2(f) = [0,\ldots,\underset{i}{1},\underset{i+1}{1},\ldots,0], \qquad \forall f \in F,$$

where i, $\lceil f/n \rceil \in \{i, i+1\}$. Then

$$V(N_1^2, f) = T_i \cup T_{i+1}$$

and

$$r(N_1^2) = r(N_1^2, f) = 2r(N) = \frac{1}{n}.$$

Finally consider an approximate information operator N_1^3 defined as

$$N_1^3(f) = [0,\ldots,0], \qquad \forall f \in F.$$

Since all correct information differs from $N_1^3(f)$ by unity, all elements of F are indistinguishable. Therefore $V(N_1^3, f) = F$ and

$$r(N_1^3) = r(N_1^3, f) = 1/2.$$

From this we conclude that

$$r(N_{1,E}) = r(N_{1,E}, f) = 1/2.$$

Thus, the radius of worst-case approximate information is independent of n. Therefore no matter how many questions we ask, we cannot solve the problem with $\epsilon < 1/2$ using any approximate information. We emphasize that this is a property of the information operator N; in Chapter 6 we exhibit an information operator which is robust under approximate information.

Finally, this example shows that the lower and upper bounds on the radius of approximate information given by (3.19) are sharp and $r(N_{\rho,E})$ may differ significantly from $r(N_\rho)$ for some approximate information N_ρ. This example will be continued in Section 2.4. ■

Example: Integration (continued) Let N_ρ be an approximate information operator with $\rho \in (0,1)$. Let $a \in \mathbb{R}_+$ and $f(x) \equiv a$. Then $f \in F$. Let $N_\rho(f) = [a,\ldots,a]$.

Define $\tilde{f}(x) = a(1 + \rho) + \rho\eta$. Then

$$N_\rho(f) \in E(N(\tilde{f}), \rho).$$

Since $\int_0^1 [\tilde{f}(x) - f(x)]dx = \rho(a + \eta)$ and a is arbitrary, we get

$$r(N_{\rho,E}) = r(N_\rho) = r(N_\rho,f) = +\infty.$$

Thus, it is impossible to find an ϵ-approximation using the approximate information N_ρ no matter how large ϵ is. One may observe that since the error in $N(f)$ is measured in the relative sense, we should redefine our problem as

$$S_1(f,\epsilon) = \{g \in \mathbb{R}: |g - \int_0^1 f(x)\,dx| \leq \epsilon(|g| + \eta)\},$$

i.e., an ϵ-approximation is also defined in terms of the relative error. One may check that any approximate information N_ρ, $\rho < 1$, now provides a finite radius. This example will be continued in Chapter 4. ■

2.4 UPPER BOUNDS

As in Section 1.4 of Chapter 1, we now derive algorithms with minimal local and global errors. Let N_ρ be an approximate information operator. An algorithm ϕ, $\phi \in \Phi(N_\rho)$, is called a *strongly optimal error algorithm* iff

$$e(\phi,N_\rho,f) = r(N_\rho,f), \qquad \forall f \in F, \tag{4.1}$$

and is called an *optimal error algorithm* iff

$$e(\phi,N_\rho) = r(N_\rho). \tag{4.2}$$

Note that for $\rho = 0$, (4.1) and (4.2) coincide with (4.1) and (4.2) of Chapter 1. Optimal and strongly optimal is with respect to a particular N_ρ.

We generalize the concept of central algorithm introduced by (4.4) of Chapter 1. Observe that (3.4) implies that $A(N_\rho,f,r(N_\rho,f) + \delta)$ is nonempty for all positive δ. Suppose there exists an element $c(f)$ such that

$$c(f) \in A(N_\rho,f,r(N_\rho,f) + \delta), \qquad \forall \delta > 0. \tag{4.3}$$

We call $c(f)$ a *center* of f and the algorithm

$$\phi^c(N_\rho(f)) = c(f) \tag{4.4}$$

is called a *central algorithm*.

THEOREM 4.1 An algorithm ϕ is a strongly optimal error algorithm iff ϕ is central. ■

We omit the proof since it is similar to the proof of Theorem 4.1 of Chapter 1.

We now discuss algorithms with minimal local and global errors from the class $\Phi(N_{\rho,E})$. An algorithm ϕ, $\phi \in \Phi(N_{\rho,E})$, is called a *strongly optimal error algorithm for a worst case* iff

$$e(\phi,N_{\rho,E},f) = r(N_{\rho,E},f), \qquad \forall f \in F, \tag{4.5}$$

and is called an *optimal error algorithm for a worst case* iff

$$e(\phi,N_{\rho,E}) = r(N_{\rho,E}). \tag{4.6}$$

We now exhibit relations between strongly optimal error algorithms and central algorithms.

THEOREM 4.2

(i) If ϕ is a central algorithm for *every* approximate information operator N_ρ, then ϕ is a strongly optimal error algorithm for a worst case.

(ii) Suppose that for every $f \in F$ the supremum in (3.13) is attained for N_ρ^*, that is,

$$N_\rho^*(f) \in E(N(f),\rho) \qquad \text{and} \qquad r(N_\rho^*,f) = r(N_{\rho,E},f), \qquad \forall f \in F.$$

If ϕ is a strongly optimal error algorithm for a worst case, then ϕ is a central algorithm for N_ρ^*.

(iii) If ϕ is an optimal error algorithm for *every* approximate information operator N_ρ, then ϕ is an optimal error algorithm for a worst case.

(iv) Suppose that the supremum in (3.16) is attained for N_ρ^*, that is,

$$N_\rho^*(f) \in E(N(f),\rho), \qquad \forall f \in F, \qquad \text{and } r(N_\rho^*) = r(N_{\rho,E}).$$

If ϕ is an optimal error algorithm for a worst case, then ϕ is an optimal error algorithm for N_ρ^*. ∎

PROOF We prove (i): If ϕ is central for N_ρ, then

$$e(\phi,N_\rho,f) = r(N_\rho,f), \qquad \forall N_\rho, \qquad \forall f \in F.$$

From (3.13) and (3.21) we have

$$\begin{aligned} r(N_{\rho,E},f) &= \sup_{N_\rho(f)\in E(N(f),\rho)} r(N_\rho,f) \\ &= \sup_{N_\rho(f)\in E(N(f),\rho)} e(\phi,N_\rho,f) \\ &= e(\phi,N_{\rho,E},f), \qquad \forall f \in F. \end{aligned}$$

This shows that ϕ is a strongly optimal error algorithm for a worst case.

We prove (ii): If ϕ is strongly optimal for a worst case then

$$e(\phi,N_{\rho,E},f) = r(N_{\rho,E},f) = r(N^*_\rho,f).$$

From (3.21),

$$e(\phi,N^*_\rho,f) \le e(\phi,N_{\rho,E},f).$$

Since $e(\phi,N^*_\rho,f) \ge r(N^*_\rho,f)$ due to (3.11), we conclude

$$e(\phi,N^*_\rho,f) = r(N^*_\rho,f), \qquad \forall f \in F,$$

which means that ϕ is a strongly optimal error algorithm for N^*_ρ.

We prove (iii): If ϕ is an optimal error algorithm for every N_ρ, then

$$e(\phi,N_\rho) = r(N_\rho),\ \forall N_\rho.$$

Taking the supremum over all N_ρ, we have from (3.22) and (3.16)

$$e(\phi,N_{\rho,E}) = r(N_{\rho,E}),$$

which means that ϕ is an optimal error algorithm for a worst case.

We finally prove (iv): If ϕ is an optimal error algorithm for a worst case, then

$$e(\psi,N_{\rho,E}) = r(N_{\rho,E}) - r(N^*_\rho).$$

From (3.22),

$$e(\phi,N^*_\rho) \le e(\phi,N_{\rho,E}).$$

Since $e(\phi,N^*_\rho) \ge r(N^*_\rho)$ due to (3.12), we conclude

$$e(\phi,N^*_\rho) = r(N^*_\rho),$$

which means that ϕ is an optimal error algorithm for N^*_ρ. ■

We illustrate the concept of central algorithm by an example.

Example: Binary Search (continued) As in Section 2.3, let

$$N_1^1(f) = [0,\ldots,\underset{j}{1},\ldots,0].$$

Since N_1^1 is correct, the algorithms

$$\phi_a(N_1^1(f)) = (j - 1)p + \tfrac{p}{2} + (a - 1), \qquad a = 1,2,$$

are central for N_1^1. Consider next

$$N_1^2(f) = [0,\ldots, \underset{i}{1}, \underset{i+1}{1},\ldots, 0].$$

Then it is easy to check that the algorithms

$$\phi_a(N_1^2(f)) = ip + (a - 1), \qquad a = 1,2,$$

are central for N_1^2. Finally, let

$$N_1^3(f) = [0,\ldots, 0].$$

Then the algorithms

$$\phi_a(N_1^3(f)) = \tfrac{m}{2} + (a - 1), \qquad a = 1,2,$$

are central for N_1^3. This example will be continued in Chapter 4. ■

The classes of interpolatory and interior algorithms that use approximate information are discussed in Appendixes A and B. See also Appendix H where further relations between central algorithms and the radius of approximate information are exhibited.

CHAPTER THREE

REALIZABLE ALGORITHMS

3.1 INTRODUCTION

In the first two chapters our definition of algorithm was extremely general. An (idealized) algorithm is any operator on the set $N_\rho(F)$. This generality strengthens the results by limiting the extent to which any algorithm can reduce uncertainty. In computational practice we often restrict the class of algorithms. We formalize this restriction by introducing the concept of realizable algorithm. It is intuitively clear that restricting the class of algorithms can only increase uncertainty. The related concept of permissible algorithm is discussed in Chapter 5.

Numerous classes of realizable algorithms occur in computational practice. They include:

1. The class of algorithms that are linear functions of the input.
2. The class of on-line algorithms.
3. The class of algorithms that are polynomials or rational functions with coefficients depending on information.

Many of our examples and applications involve a class of realizable algorithms. They include:

1. Polynomial zeros with algorithms constructed from arithmetic operations or arithmetics plus radicals (Section 3.2).
2. Uniform approximation with algorithms constructed from $+$, $-$, $*$ (Section 3.2).
3. Division with algorithms constructed from $+$, $-$, $*$ (Section 5.4 of Chapter 5).
4. Bin packing with on-line algorithms (Section 6.5 of Chapter 6).

We summarize the results of this chapter. In Section 3.2 we define the class of realizable algorithms. We introduce polynomial zeros and uniform approximation examples to illustrate several classes of realizable algorithms. In Sections 3.3 and 3.4 we extend the results of the corresponding sections of Chapter 2 to realizable algorithms.

3.2 REALIZABLE ALGORITHMS

Let N_ρ be an approximate information operator, $N_\rho : F \to H$. Recall that $\Phi(N_\rho)$ is the class of idealized algorithms ϕ, ϕ: $N_\rho(F) \to G$, that use N_ρ. It is often necessary to impose some restrictions on ϕ in order to guarantee that the element $\phi(N_\rho(f))$ can be easily computed and/or ϕ enjoys some useful properties.

Let R be a subset of $\phi(N_\rho)$. We say that R is a class of *realizable algorithms* and ϕ, $\phi \in R$, is a *realizable algorithm*. We illustrate this concept by a number of examples.

Example: Polynomial Zeros Let F be the class of all polynomials of degree n. Let $G = \mathbb{C}$ and define

$$S(f,\epsilon) = \{\alpha \in G: f(\alpha) = 0\}.$$

Thus $x \in S(f,\epsilon)$ means that x is a zero of f. Let $\rho = 0$ and $H = N(F)$ with $N(f) = [a_0,a_1,\ldots,a_n]$, where $f(x) = a_0 + a_1x + \cdots + a_nx^n$, $a_n \neq 0$. For a given integer k, $k \geq 2$, define

$$R = R_1 = \{\phi\text{: the evaluation of } \phi(a_0,a_1,\ldots,a_n) \text{ requires at most } k$$
$$\text{arithmetic operations, } \forall (a_0,\ldots,a_n) \in H\}.$$

Thus, if $\phi \in R_1$, then ϕ is a rational function and the cost of computing $\phi(a_0,\ldots,a_n)$ is at most the cost of performing k arithmetic operations. Define a second class of realizable algorithms as

$$R = R_2 = \{\phi\text{: the evaluation of } \phi(a_0,\ldots,a_n) \text{ requires at most } k$$
$$\text{arithmetic operations or radicals, } \forall (a_0,\ldots,a_n) \in H\}.$$

Thus, we now allow radicals as well as arithmetic operations. This example will be continued in Section 3.3. ■

Example: Uniform Approximation Let $G = C([0,1])$ be the class of real continuous functions on [0,1]. Let $\| f \| = \max_{x \in [0,1]} |f(x)|$. Define $F = \{f \in G : \|f\| \leq 1\}$ and

$$S(f,\epsilon) = \{g \in G : \| f - g \| \leq \epsilon\}.$$

Let $H = G$ and $E(h,\rho) = \{g \in H : \| g - h \| \leq \rho\}$. Let $N(f) = f$. Then

$\| N_\rho(f) - f \| \le \rho$. Define $R = \{\phi: \phi(h) \in \pi_n, \forall h \in N_\rho(F)\}$, where π_n is the class of all polynomials of degree at most n. Thus the class R of realizable algorithms is the class of all algorithms ϕ such that $\phi(N_\rho(f))$ is a polynomial of degree at most n, that is,

$$\phi(N_\rho(f))(x) = a_0 + a_1 x + \cdots + a_n x^n$$

with $a_i = a_i(N_\rho(f))$, $i = 0,1,\ldots,n$.

This is the formulation of the uniform approximation problem in our terminology. This example will be continued in Section 3.3. ■

3.3 LOWER BOUNDS

We extend the results of Chapters 1 and 2 for the class R of realizable algorithms. As always, we seek an ϵ-approximation, i.e., an element x which belongs to

$$A(N_\rho,f,\epsilon) = \bigcap_{\tilde{f} \in V(N_\rho,f)} S(\tilde{f},\epsilon).$$

We now require that $x = \phi(N_\rho(f))$ for some realizable algorithm ϕ. Let $h = N_\rho(f)$ and

$$Q(h) = \{\phi(h) : \phi \in R\} \tag{3.1}$$

be the set of values of all realizable algorithms at h. Then x belongs to $A(N_\rho,f,\epsilon) \cap Q(h)$. Thus we can find an ϵ-approximation iff $A(N_\rho,f,\epsilon) \cap Q(h)$ is nonempty. Otherwise, if $A(N_\rho,f,\epsilon) \cap Q(h) = \emptyset$, our "adversary" can select a set $S(\tilde{f},\epsilon)$ to which x does not belong.

We now generalize the concept of local radius for R. Let

$$r(R,N_\rho,f) = \inf\{\delta: A(N_\rho,f,\delta) \cap Q(N_\rho(f)) \ne \emptyset\} \tag{3.2}$$

be the *local radius of approximate information* (for R). Note that

$$r(R,N_\rho,f) \ge r(\Phi(N_\rho),N_\rho,f) = r(N_\rho,f). \tag{3.3}$$

Furthermore, if $Q(N_\rho(f)) = G$, then

$$r(R,N_\rho,f) = r(N_\rho,f). \tag{3.4}$$

We now generalize the concept of (global) radius for R. For a given δ, let

$$\Phi(\delta) = \{\phi \in \Phi(N_\rho) : \phi(N_\rho(f)) \in A(N_\rho,f,\delta) \cap Q(N_\rho(f)), \forall f \in F\} \tag{3.5}$$

be the set of all algorithms from $\Phi(N_\rho)$ whose values belong to the sets $A(N_\rho,f,\delta)$

$\cap\ Q(N_\rho(f))$. Note that ϕ from $\Phi(\delta)$ need not be an element of R. We wish to find the minimal value of δ for which $\Phi(\delta)$ contains an algorithm from R. More precisely, let

$$r(R,N_\rho) = \inf\{\delta\colon \Phi(\delta) \cap R \neq \emptyset\} \tag{3.6}$$

be the (global) *radius of approximate information* (for R). Note that (3.2) yields

$$\inf\{\delta\colon \Phi(\delta) \neq \emptyset\} = \sup_{f \in F} r(R,N_\rho,f).$$

Thus

$$r(R,N_\rho) \geq \sup_{f \in F} r(R,N_\rho,f). \tag{3.7}$$

In contrast to Chapters 1 and 2, the strict inequality in (3.7) can occur, as is shown by the following example.

Example: Polynomial Zeros (continued) We calculate the radii $r(R_i,N,f)$ and $r(R_i,N)$ for $i = 1, 2$. First, note that $Q(h) = G$ for both R_i. Indeed, let $g \in G$ and define $\phi(a_0,\ldots,a_n) \equiv g$. Since the evaluation of ϕ requires no arithmetic operations or radicals, $\phi \in R_i$ and $g \in Q(h)$. This implies $Q(h) = G$. Thus

$$r(R_i,N,f) = 0, \qquad \forall f \in F,$$

which means that for every f there exists a realizable algorithm ϕ with $e(\phi,f) = 0$. Note that this ϕ depends on f.

Let $\phi \in \Phi(\delta)$. Then $\phi(a_0,\ldots,a_n)$ is a zero of the polynomial $f(x) = a_0 + a_1x + \cdots + a_nx^n$. On the other hand, if $\phi \in R_1$, then the evaluation of $\phi(a_0,\ldots,a_n)$ requires at most k arithmetic operations, $k \geq 2$. Therefore $\Phi(\delta) \cap R_1$ is nonempty iff $n = 1$. Thus

$$r(R_1,N) = \begin{cases} 0 & \text{if } n = 1, \\ +\infty & \text{otherwise.} \end{cases}$$

For R_2 we can also compute radicals. From Abel's theorem we know that these operations permit us to compute a zero of a general polynomial iff its degree is at most 4. Thus, provided we can perform enough arithmetic operations and radicals (i.e., k is large enough) we have

$$r(R_2,N) = \begin{cases} 0 & \text{if } n \leq 4, \\ +\infty & \text{otherwise.} \end{cases}$$

This shows that for this example the left-hand side of (3.7) is infinity, whereas the right-hand side is zero. ■

We now present a condition that guarantees equality in (3.7). Let

$$\bar{R} = \{\phi \in \Phi(N_\rho) : \phi(N_\rho(f)) \in Q(N_\rho(f)),\ \forall f \in F\}. \tag{3.8}$$

Obviously $R \subset \bar{R}$. We now prove

LEMMA 3.1 If $R = \bar{R}$, then

$$r(R,N_\rho) = \sup_{f \in F} r(R,N_\rho,f). \tag{3.9}$$

■

PROOF Let

$$a = \sup_{f \in F} r(R,N_\rho,f).$$

If $a = +\infty$, then (3.9) follows from (3.7). Assume then that $a < +\infty$. Let $\delta > a$. From (3.2) we know that $A(N_\rho,f,\delta) \cap Q(N_\rho(f))$ is nonempty for every f. Let $x = x(N_\rho(f))$ be an element of this set. Define $\phi(N_\rho(f)) = x(N_\rho(f))$, $\forall f \in F$. Then $\phi \in \bar{R}$ and $\phi \in \Phi(\delta)$. Since $\bar{R} = R$, we conclude that $\Phi(\delta) \cap R$ is nonempty for any $\delta > a$. This yields $r(R,N_\rho) \le a$ and (3.9) follows from (3.7). ■

We now provide an example for which $R = \bar{R}$.

Example 3.1 Let Q be a subset of G and define

$$R = \{\phi: \phi(h) \in Q,\ \forall h \in N_\rho(F)\}.$$

We prove that $R = \bar{R}$. Indeed, let $h = N_\rho(f)$. We first show that $Q(h) = Q$. Let $q \in Q$ and defining $\phi(a) \equiv q$, $\forall a \in N_\rho(F)$, we see that $\phi \in R$ and $q \in Q(h)$. Since q is arbitrary, $Q \subset Q(h)$. From the definition of $Q(h)$ we have $Q(h) \subset Q$, which yields $Q(h) = Q$. To show $R = \bar{R}$, observe that $\phi \in \bar{R}$ implies $\phi(h) \in Q(h) = Q$. Thus $\phi \in R$ and $\bar{R} \subset R$. Hence $R = \bar{R}$ as claimed.

Note that for $Q = G$ we get $R = \bar{R} = \Phi(N_\rho)$. Of course,

$$r(R,N_\rho) \ge r(\Phi(N_\rho),N_\rho) = r(N_\rho) = \sup_{f \in F} r(N_\rho,f),$$

which agrees with Chapters 1 and 2. ■

We now illustrate the computation of radii for the uniform approximation problem and the normed linear space setting.

Example: Uniform Approximation (continued) We compute the radius $r(R,N_\rho,f)$, where $R = \Pi_n$. Note that $V(N_\rho,f) = \{\tilde{f}: \| \tilde{f} - N_\rho(f) \| \le \rho\}$ is a ball with center $N_\rho(f)$ and radius ρ. Let

$$\operatorname{dist}(\Pi_n,N_\rho(f)) = \inf_{g \in \pi_n} \| N_\rho(f) - g \| = \| N_\rho(f) - z \| ,$$

where $z \in \Pi_n$. Define $\delta = \rho + \operatorname{dist}(\Pi_n,N_\rho(f))$. For any $\tilde{f}$ from $V(N_\rho,f)$ we have

$$\| \tilde{f} - z \| \le \| \tilde{f} - N_\rho(f) \| + \| N_\rho(f) - z \| \le \delta.$$

Thus $z \in A(N_\rho,f,\delta) = \bigcap_{\tilde{f} \in V(N_\rho,f)} \{g: \| \tilde{f} - g \| \le \delta\}$. Since $Q(N_\rho(f)) = \Pi_n$, $\forall f \in F$, we have $r(R,N_\rho,f) \le \delta$. Let $\| N_\rho(f) - z \| > 0$. Define

$$\tilde{f} = N_\rho(f) + \frac{\rho}{\| N_\rho(f) - z \|} (N_\rho(f) - z);$$

then $\tilde{f} \in V(N_\rho,f)$. Take any $g \in \Pi_n$, then

$$\| \tilde{f} - g \| = \left[1 + \frac{\rho}{\| N_\rho(f) - z \|}\right] \| N_\rho(f) - g_1 \| \ge \delta,$$

where $g_1 \in \Pi_n$. This shows that $r(R,N_\rho,f) \ge \delta$.

Suppose now that $N_\rho(f) = z$. It is obvious that there exists $f_1 \in F$ such that $\operatorname{dist}(f_1,\Pi_n) = 1$. Then $\| f_1 \| = 1$. Define $\tilde{f} = N_\rho(f) + \rho f_1$. We have $\tilde{f} \in V(N_\rho,f)$ and

$$\| \tilde{f} - g \| \ge \rho, \qquad \forall g \in \Pi_n.$$

This proves that $r(R,N_\rho,f) \ge \rho$. Combining all cases, we conclude that

$$r(R,N_\rho,f) = \rho + \operatorname{dist}(\Pi_n, N_\rho(f)).$$

We now compute $r(R,N_\rho)$. Setting $Q = \Pi_n$ in Example 3.1 and applying Lemma 3.1, we conclude that

$$r(R,N_\rho) = \rho + \sup_{f \in F} \operatorname{dist}(\Pi_n,N_\rho(f)).$$

Due to $\| f \| \le 1$, we have $\| N_\rho(f) \| \le 1 + \rho$ and $\operatorname{dist}(\Pi_n,N_\rho(f)) \le 1 + \rho$. Thus

$$r(R,N_\rho) \le 1 + 2\rho.$$

Note that for $g \in \Pi_n$ we have

$$\| N_\rho(f) - g \| \geq \| f - g \| - \| N_\rho(f) - f \| \geq \operatorname{dist}(\Pi_n, f) - \rho.$$

Choose f such that $\operatorname{dist}(\Pi_n, f) = 1$. For this f, $\operatorname{dist}(\Pi_n, N_\rho(f)) \geq \max(1 - \rho, 0)$. This yields

$$\max(1,\rho) \leq r(R,N_\rho) \leq 1 + 2\rho, \qquad \forall n.$$

These two bounds are sharp, i.e., by setting $N_\rho(f) = cf$, where $c = 1 - \rho$ for $\rho \leq 1$ and $c = 0$ for $\rho \geq 1$, we get the left-hand side, and by setting $N_\rho(f) = (1 + \rho)f$ we get the right-hand side. This example will be continued in this section. ■

Example: Normed Linear Space (continued) Let Q be a subset of G and $R = \{\phi\colon \phi(h) \in Q, \forall h \in N_\rho(F)\}$. Then $Q(h) \equiv Q$ and $R = \bar{R}$. For two given subsets P_1 and P_2 of G we define

$$\operatorname{dist}(P_1,P_2) = \inf_{a \in P_1} \inf_{b \in P_2} \|a - b\|,$$

$$\operatorname{diam}(P_1) = \sup_{a_1,a_2 \in P_1} \|a_1 - a_2\|.$$

We prove

LEMMA 3.2 Let $\rho = 0$. Let $P = S(V(N,f))$. Then

$$\max(\operatorname{dist}(P,Q), \operatorname{diam}(P)/2) \leq r(R,N,f) \leq \operatorname{dist}(P,Q) + \operatorname{diam}(P), \tag{3.10}$$

$$r(R,N,f) = c(\operatorname{dist}(P,Q) + \operatorname{diam}(P)), \tag{3.11}$$

where $c \in [1/3,1]$. ■

PROOF Recall that in this case

$$r = r(R,N,f) = \inf\{\delta\colon A(N,f,\delta) \cap Q \neq \emptyset\}$$

with

$$A(N,f,\delta) = \bigcap_{\tilde{f} \in V(N,f)} \{g\colon \|S(\tilde{f}) - g\| < \delta\} \text{ for } \delta > 0.$$

Assume that r is finite and take $\delta > r$. Then there exists an element g of Q such that

$$\operatorname{dist}(P,Q) \leq \| S(\tilde{f}) - g \| < \delta, \qquad \forall \tilde{f} \in V(N,f). \tag{3.12}$$

For $f_1,f_2 \in V(N,f)$ we have

$$\| S(f_1) - S(f_2) \| \le \| S(f_1) - g \| + \| S(f_2) - g \| < 2\delta. \tag{3.13}$$

Letting δ tend to r in (3.12) and (3.13) and taking the supremum with respect to f_1 and f_2 in (3.13), we get

$$\max(\mathrm{dist}(P,Q),\ \mathrm{diam}(P)/2) \le r.$$

This proves the left-hand side of (3.10). If $r = +\infty$, the left-hand side of (3.10) holds trivially.

We now prove the right-hand side of (3.10). Let $\eta > 0$. Take $g \in Q$ and $S(\tilde{f})$ from P such that

$$\| S(\tilde{f}) - g \| \le \mathrm{dist}(P,Q) + \eta.$$

Let $f_1 \in V(N,f)$. Then

$$\| S(f_1) - g \| \le \| S(f_1) - S(\tilde{f}) \| + \| S(\tilde{f}) - g \| \le \mathrm{diam}(P) + \mathrm{dist}(P,Q) + \eta.$$

Thus $g \in A(N,f, \mathrm{diam}(P) + \mathrm{dist}(P,Q) + \eta) \cap Q$. Since η is arbitrary, this proves that

$$r \le \mathrm{dist}(P,Q) + \mathrm{diam}(P).$$

Hence (3.10) holds. To prove (3.11) observe that

$$(a + b)/3 \le \max(a, b/2)$$

for any nonnegative numbers a and b. This completes the proof of Lemma 3.2. This example will be continued in Section 4.3 of Chapter 4. ■

We now show that $r(R,N_\rho,f)$ and $r(R,N_\rho)$ are the sharp lower bounds on the local and global errors of any algorithm from R. Recall that the local and global errors of an algorithm are defined by (3.9) and (3.10) of Chapter 2.

THEOREM 3.1

$$\inf_{\phi \in R} e(\phi,N_\rho,f) = r(R,N_\rho,f), \qquad \forall f \in F, \tag{3.14}$$

$$\inf_{\phi \in R} e(\phi,N_\rho) = r(R,N_\rho). \tag{3.15}$$

■

PROOF Let $f \in F$, $h = N_\rho(f)$ and ϕ be an algorithm from R. Then $\phi(h) \in Q(h)$ due to (3.1). Without loss of generality, assume that $a = e(\phi,N_\rho,f) < +\infty$. From (3.9) of Chapter 2 we conclude that $\phi(h) \in A(N_\rho,f,a+\delta)$, $\forall\delta > 0$. Thus $A(N_\rho,f,a+\delta) \cap Q(h)$ is nonempty and (3.2) yields $r = r(R,N_\rho,f) \leq e(\phi,N_\rho,f)$. Since ϕ is arbitrary, we get $r \leq \inf\{e(\phi,N_\rho,f)\colon \phi \in R\}$.

To prove equality, take any positive δ and assume without loss of generality that $r < +\infty$. Observe that $A(N_\rho,f,r+\delta) \cap Q(h) \neq \emptyset$ due to (3.2). Let $a(h)$ be an element of this intersection. From (3.1) it follows that there exists an algorithm ϕ from R such that

$$\phi(h) = a(h) \in A(N_\rho,f,r+\delta) \cap Q(h).$$

Thus $e(\phi,N_\rho,f) \leq r + \delta$. Since δ is arbitrary, this proves that $\inf\{e(\phi,N_\rho,f) : \phi \in R\} \leq r$. Hence (3.14) is proven.

We now prove (3.15). Let $\phi \in R$. Without loss of generality, assume that $e(\phi,N_\rho) < +\infty$. Then $\phi(h) \in Q(h)$, $h = N_\rho(f)$, and $\phi(h) \in A(N_\rho,f,\delta)$, $\forall f \in F$, for any $\delta > e(\phi,N_\rho)$. Thus $\phi \in \Phi(\delta) \cap R$ and (3.6) yields $r(R,N_\rho) < \delta$. Letting δ tend to $e(\phi,N_\rho)$, we get $r(R,N_\rho) \leq e(\phi,N_\rho)$. Since ϕ is arbitrary, $r(R,N_\rho) \leq \inf\{e(\phi,N_\rho) : \phi \in R\}$.

To prove equality, we can assume that $r(R,N_\rho)$ is finite. Take $\delta > r(R,N_\rho)$. Then $\Phi(\delta) \cap R \neq \emptyset$, which means that there exists an algorithm ϕ such that $\phi \in R$ and $\phi(N_\rho(f)) \in A(N_\rho,f,\delta)$, $\forall f \in F$. Thus $e(\phi,N_\rho) < \delta$ and, letting δ tend to $r(R,N_\rho)$, we have $\inf\{e(\phi,N_\rho)\colon \phi \in R\} \leq r(R,N_\rho)$. Hence (3.15) is proven. ■

From Theorem 3.1 we immediately obtain

COROLLARY 3.1 It is possible to find an ϵ-approximation for all $f \in F$ by using an approximate information N_ρ and an algorithm from R iff

(i) $r(R,N_\rho) < \epsilon$, or

(ii) $r(R,N_\rho) = \epsilon$ and the infimum of (3.15) is attained for an algorithm ϕ and for this algorithm the infimum of (3.10) of Chapter 2 is also attained.

■

As in Section 2.3 of Chapter 2, we investigate the radius of approximate information in terms of $E(N(f),\rho)$ instead of N_ρ. For simplicity we assume that the set $N_\rho(F)$ is constant for any approximate information N_ρ, $N_\rho(F) \equiv H_1$. Then the class R of algorithms $\phi\colon N_\rho(F) = H_1 \to G$ can be treated as a subset of $\Phi(N_\rho)$ for any approximate information N_ρ. Let

$$r(R,N_{\rho,E},f) = \sup_{N_\rho\colon N_\rho(f) \in E(N(f),\rho)} r(R,N_\rho,f) \tag{3.16}$$

be the *local radius of worst case approximate information* (for R). Obviously,

$$r(R,N_{\rho,E},f) \geq r(\Phi(N_{\rho,E}), N_{\rho,E},f) = r(N_{\rho,E},f), \tag{3.17}$$

$$\rho_1 \leq \rho_2 \text{ implies } r(R,N_{\rho_1,E},f) \leq r(R,N_{\rho_2,E},f), \qquad \forall f \in F. \tag{3.18}$$

Similarly, let

$$r(R,N_{\rho,E}) = \sup_{N_\rho:\ N_\rho(f) \in E(N(f),\rho),\ \forall f \in F} r(R,N_\rho) \tag{3.19}$$

be the (global) *radius of worst case information* (for R). Obviously,

$$r(R,N_{\rho,E}) \geq r(\Phi(N_{\rho,E}), N_{\rho,E}) = r(N_{\rho,E}), \tag{3.20}$$

$$\rho_1 \leq \rho_2 \text{ implies } r(R,N_{\rho_1,E}) \leq r(R,N_{\rho_2,E}). \tag{3.21}$$

For fixed ρ we have

$$r(R,N) = r(R,N_{0,E}) \leq r(R,N_\rho) \leq r(R,N_{\rho,E}) \tag{3.22}$$

for any approximate information operator N_ρ.

We illustrate the computation of this global radius.

Example: Uniform Approximation (continued) Since $r(R,N_\rho) \leq 1 + 2\rho$ and this bound is sharp, we get

$$r(R,N_{\rho,E}) = 1 + 2\rho, \qquad \forall n.$$

This example will be continued in Section 3.4. ■

We now show that $r(R,N_{\rho,E},f)$ and $r(R,N_{\rho,E})$ are the sharp lower bounds on the local and global errors of an algorithm from the class R (see (3.21) and (3.22) of Chapter 2 for the definitions of local and global error).

THEOREM 3.2

$$\inf_{\phi \in R} e(\phi,N_{\rho,E},f) = r(R,N_{\rho,E},f), \qquad \forall f \in F, \tag{3.23}$$

$$\inf_{\phi \in R} e(\phi,N_{\rho,E}) = r(R,N_{\rho,E}). \tag{3.24}$$

■

The proof is straightforward and similar to the proof of Theorem 3.1 and therefore is omitted.

From Theorem 3.2 we conclude

COROLLARY 3.2 It is possible to find an ϵ-approximation for all $f \in F$ by using *any* approximate information N_ρ and an algorithm from R iff

(i) $r(R,N_{\rho,E}) < \epsilon$, or

(ii) $r(R,N_{\rho,E}) = \epsilon$ and the supremum in (3.19) or (3.16) is *not* attained, or

(iii) $r(R,N_{\rho,E}) = \epsilon$ and the suprema in (3.19) and (3.16) are attained for N_ρ for which (ii) of Corollary 3.1 holds. ∎

Examples of classes of realizable algorithms are also given in Chapter 6.

3.4 UPPER BOUNDS

As before, we derive algorithms with minimal local and global error. Let N_ρ be an approximate information operator and R a class of realizable algorithms. An algorithm ϕ, $\phi \in R$, is called a *strongly optimal error algorithm* (for R) iff

$$e(\phi,N_\rho,f) = r(R,N_\rho,f), \qquad \forall f \in F, \tag{4.1}$$

and is called an *optimal error algorithm* (for R) iff

$$e(\phi,N_\rho) = r(R,N_\rho). \tag{4.2}$$

Optimal and strongly optimal is with respect to a particular N_ρ and R. Note that if $Q(N_\rho(f)) = G$, $\forall f \in F$, and $R = \bar{R}$, then (4.1) and (4.2) coincide with (4.1) and (4.2) of Chapter 2.

We define a central algorithm as before. That is, observe first that $A(N_\rho,f,r(R,N_\rho,f) + \delta) \cap Q(N_\rho(f))$ is nonempty for all positive δ and all $f \in F$. Suppose that for every $f \in F$ there exists an element $c(f)$ such that

$$c(f) \in A(N_\rho,f,r(R,N_\rho,f) + \delta) \cap Q(N_\rho(f)), \qquad \forall \delta > 0. \tag{4.3}$$

We call $c(f)$ a *center of* f in $Q(N_\rho(f))$, and the algorithm

$$\phi^c(N_\rho(f)) = c(f) \tag{4.4}$$

is called a *central algorithm*. Note that ϕ^c need not belong to R.

THEOREM 4.1 If an algorithm ϕ is a strongly optimal error algorithm, then ϕ is central.

If a central algorithm ϕ belongs to R, then ϕ is a strongly optimal error algorithm. ∎

The proof is omitted since it is obvious.

We illustrate a central algorithm by an example.

Example: Uniform Approximation (continued) Let $p_n = p_n(N_\rho(f))$ be the best uniform approximation polynomial to $N_\rho(f)$, that is,

$$\inf_{g \in \Pi_n} \| N_\rho(f) - g \| = \| N_\rho(f) - p_n \|.$$

It is known that p_n always exists and is unique. It is obvious that p_n is a unique center of f in Π_n and therefore

$$\phi(N_\rho(f)) = p_n$$

is a unique central algorithm. Since $\phi \in R$, then ϕ is a strongly optimal algorithm. ■

We now discuss algorithms from the class R with minimal local and global errors. An algorithm $\phi, \phi \in R$, is called a *strongly optimal error algorithm for a worst case* (for R) iff

$$e(\phi, N_{\rho,E}, f) = r(R, N_{\rho,E}, f), \qquad \forall f \in F, \tag{4.5}$$

and is called an *optimal error algorithm for a worst case* (for R) iff

$$e(\phi, N_{\rho,E}) = r(R, N_{\rho,E}). \tag{4.6}$$

Based on Theorem 4.2 of Chapter 2, it is straightforward to prove

THEOREM 4.2

(i) If ϕ, $\phi \in R$, is a central algorithm for *every* approximate information operator N_ρ and R, then ϕ is a strongly optimal error algorithm for a worst case.

(ii) Suppose that for every $f \in F$ the supremum in (3.16) is attained for N_ρ^*, that is, $N_\rho^*(f) \in E(N(f),\rho)$ and

$$r(R, N_\rho^*, f) = r(R, N_{\rho,E}, f), \qquad \forall f \in F.$$

If ϕ is a strongly optimal error algorithm for a worst case, then ϕ is a central algorithm for N_ρ^* and R.

(iii) If ϕ is an optimal error algorithm for any approximate information operator N_ρ and R, then ϕ is an optimal error algorithm for a worst case.

(iv) Suppose that the supremum in (3.19) is attained for N_ρ^*, that is,

$$N_\rho^*(f) \in E(N(f),\rho), \qquad \forall f \in F, \qquad \text{and} \qquad r(R,N_\rho^*) = r(R,N_{\rho,E}).$$

If ϕ is an optimal error algorithm for a worst case, then ϕ is an optimal error algorithm for N_ρ^* and R. ■

Further relations between central algorithms in R and the radius of approximate information in R are exhibited in Appendix H. The class of interpolatory algorithms in R that use approximate information is discussed in Appendix A.

CHAPTER FOUR

OPTIMAL INFORMATION

4.1 INTRODUCTION

Until now, the information N was regarded as fixed. We now vary the information and pose the question of determining "optimal" information. To study optimal information we must introduce the concept of a unit of information. We call the unit a *primitive information operator.* If N consists of n primitive information operators, we say that n is the *cardinality of N.*

A basic question is when "adaptive" information is more powerful than "nonadaptive" information. *Nonadaptive* means that all the primitive information operators are mutually independent. *Adaptive* means that the choices of the primitive information may be made sequentially. Note that in a parallel model of computation with n processors, nonadaptive information of cardinality n can be computed in unit time, whereas adaptive information of the same cardinality requires time proportion to n.

Is adaptive information "stronger" than nonadaptive? We provide two very general conditions under which it is not more powerful. There are of course many problems for which adaptive information is far more powerful. Examples are provided in subsection (i), Section 6.6 of Chapter 6, and in GTOA, Part A, Chapter 8.

We summarize the results of Chapter 4. In Section 4.2 we define the fundamental concepts of primitive information operator, cardinality of information, adaptive and nonadaptive information, nth optimal information, and ϵ-cardinality number. We show that nonadaptive information is as powerful as adaptive information for the binary search and integration examples.

In Section 4.3 we present two general conditions (Theorems 3.1 and 3.2) under which adaption does not help. One of these theorems is applied to the normed linear space setting. The concluding section extends the results of Section 4.3 to approximate information.

4.2 OPTIMAL EXACT INFORMATION

Let H_1 be a given set. Let $\Psi(H_1)$ be a class of (exact) information operators U

such that

$$U : F \to H_1. \tag{2.1}$$

We call elements of $\Psi(H_1)$ *primitive information operators*. For a fixed integer n, let $H = H_1^n$ and define the (exact) information operator N, $N : F \to H$, as

$$N(f) = [U_1(f), U_2(f),..., U_n(f)], \tag{2.2}$$

where $U_i \in \Psi(H_1)$. Thus N consists of n primitive information operators. We call n the *cardinality of* N, n = card(N). To stress the dependence on the cardinality, we sometimes write $N = N_n$.

Note that the U_i in (2.2) are given simultaneously, i.e., the choice of U_{i+1} does not depend on the previous choice of i primitive information operators, $i = 1,2,..., n - 1$. Therefore, N is called *nonadaptive* and is sometimes denoted by $N = N^{\text{non}}$.

We define *adaptive information* as follows. Let

$$N^a(f) = [U_1(f), U_2(f;y_1),..., U_n(f;y_1,..., y_{n-1})], \tag{2.3}$$

where

$$y_i = U_i(f;y_1,..., y_{i-1}), \qquad i = 1, 2,..., n - 1, \tag{2.4}$$

and $U_i(\cdot;y_1,..., y_{i-1}) \in \Psi(H_1)$.

The number n is called the *cardinality of* N^a, n = card(N^a). If N^a has cardinality n, we sometimes write $N^a = N_n^a$.

The adaptive information N^a consists of n primitive information operators, and the choice of U_i *may* depend on the previously computed $y_1,..., y_i$ for $i = 1, 2,..., n$. Note that if none of the U_i depends on $y_1,..., y_{i-1}$, then adaptive information is nonadaptive.

We illustrate nonadaptive and adaptive information by two examples.

Example: Binary Search (continued) Let $H_1 = \{0,1\}$ and $\Psi(H_1)$ be the class of all single questions, i.e.,

$$U(f) = Q(f,T) = \begin{cases} 1 & \text{if } f \in T, \\ 0 & \text{if } f \notin T, \end{cases}$$

for some set $T = T(U) \subset F$. Let

$$N(f) = [Q(f,T_1), Q(f,T_2), ..., Q(f,T_n)].$$

As in Section 1.4 of Chapter 1, let

$$T_i = \{(i-1)p+1, \ldots, (i-1)p+p\}, \qquad p = m/n.$$

Setting $U_i(f) = Q(f, T_i)$ in (2.2), we see that N is nonadaptive.

We now, for the first time, present an *adaptive information operator.* It is based on the idea of "bisection," that is, T_1 is the first half of F. If f belongs to T_1, we select the first half of T_1 to be T_2, otherwise the first half of $F - T_1$ is T_2, and so on. It comes as no surprise that this adaptive bisection information is optimal, which will be proven later in this section.

To define bisection information precisely, assume for simplicity that $m = 2^s$, $s \geq n$, and define

$$T_i^b = \{k_i + 1, k_i + 2, \ldots, k_i + 2^{s-i}\}$$

for $i = 0, 1, \ldots, n$, where $k_0 = 0$ and

$$k_{i+1} = \begin{cases} k_i & \text{if} \quad Q(f, T_i^b) = 1, \\ k_i + 2^{s-i} & \text{if} \quad Q(f, T_i^b) = 0. \end{cases}$$

Note that $T_0^b = F$, the set T_i^b consists of 2^{s-i} elements and $f \in T_i^b$ or $f \in \{k_i + 2^{s-i} + 1, \ldots, k_i + 2*2^{s-i}\}$ for $i = 1, 2, \ldots n$.

The choice of T_{i+1}^b depends on the previous answers $Q(f, T_1^b), \ldots, Q(f, T_i^b)$. Setting

$$N^a(f) = [Q(f, T_1^b), \ldots, Q(f, T_n^b)], \tag{2.5}$$

which corresponds to $y_i = Q(f, T_i^b)$ and $U_i(f; y_1, \ldots, y_{i-1}) = Q(f, T_i^b)$ in (2.3), we see that the bisection information N^a is *adaptive.* This example will be continued in this section. ■

Example: Integration (continued) Let $H_1 = \mathbb{R}$ and $\Psi(H_1)$ be the class of all single function evaluations, i.e.,

$$U(f) = f(x)$$

for some point $x = x(U) \in [0,1]$. Let

$$N(f) = [f(x_1), f(x_2), \ldots, f(x_n)].$$

If $x_i = (i-1)/(n-1)$, as in Section 1.4 of Chapter 1, then, setting $U_i(f) = f(x_i)$ in (2.2), we see that N is nonadaptive.

If the choice of x_i depends on the previously computed values $f(x_1), \ldots, f(x_{i-1})$, that is

$$x_i = x_i(f(x_1),\ldots, f(x_{i-1})) \not\equiv \text{const},$$

then N is adaptive. This example will be continued in this section. ■

Define

$$\Psi^{\text{non}}(n) = \{N^{\text{non}} : \text{card}(N^{\text{non}}) \le n\} \tag{2.6}$$

as the class of all nonadaptive information with cardinality at most n and

$$\Psi^{a}(n) = \{N^{a} : \text{card}\ (N^{a}) \le n\} \tag{2.7}$$

as the class of all adaptive information with cardinality at most n. Clearly, $\Psi^{\text{non}}(n) \subset \Psi^{a}(n)$. We emphasize that both classes $\Psi^{\text{non}}(n)$ and $\Psi^{a}(n)$ depend on the class $\Psi(H_1)$, i.e., on the choice of primitive information operators.

In Chapter 3 we denoted by R a class of realizable algorithms, $R \subset \Phi(N)$. Note that R depends on N since R contains algorithms with the domain $N(F)$. Since we vary N in this chapter, we shall always denote by $R = R(N)$ the class of realizable algorithms that use N. Thus $r(R(N),N)$ denotes the radius of information N for the class $R(N)$ of realizable algorithms, $N \in \Psi^{a}(n)$.

We define the *nth minimal radius of information* (for $\Psi^{\text{non}}(n)$ and R) or ($\Psi^{a}(n)$ and R) as

$$R^{\text{non}}(R,n) = \inf_{N\in\Psi^{\text{non}}(n)} r(R(N),N)$$

$$(\text{or } r^{a}(R,n) = \inf_{N\in\Psi^{a}(n)} r(R(N),N)). \tag{2.8}$$

Note that the symbol R above does not denote the class of realizable algorithms, but is to emphasize that we are using classes of realizable algorithms.

An information operator N is *nth optimal information* (for $\Psi^{\text{non}}(n)$ and R) or ($\Psi^{a}(n)$ and R) iff

$$N \in \Psi^{\text{non}}(n) \qquad (\text{or } N \in \Psi^{a}(n)), \tag{2.9}$$

$$r(R(N),N) = r^{\text{non}}(R,n) \qquad (\text{or } r(R(N),N) = r^{a}(R,n)). \tag{2.10}$$

Clearly $r^{a}(R,n) \le r^{\text{non}}(R,n)$. Thus, an nth optimal information has the smallest radius of information among all information with cardinality at most n. We again emphasize that the nth minimal radius and the nth optimal information depend on the choice of primitive information operators. If $R(N) = \Phi(N)$, $\forall N$, we do not list R in (2.8) and (2.10) and simply write $r^{\text{non}}(n)$, $r^{a}(n)$.

What is the minimal cardinality of information such that we can find an ϵ-approximation? From (2.8) we conclude that we have to guarantee that $r^{\text{non}}(R,n)$

or $r^a(R,n)$ does not exceed ϵ. This motivates the following definition.

We define the *ϵ-cardinality number* (for nonadaptive or adaptive information and for R) as

$$m^{\text{non}}(R,\epsilon) = \min\{n: r^{\text{non}}(R,n) \le \epsilon\}$$
$$(\text{or } m^a(R,\epsilon) = \min\{n: r^a(R,n) \le \epsilon\}). \tag{2.11}$$

If $R(N) = \Phi(N)$, $\forall N$, we do not list R in (2.11) and simply write $m^{\text{non}}(\epsilon)$ or $m^a(\epsilon)$. Thus, the ϵ-cardinality is the smallest cardinality for which the nth minimal radius of information does not exceed ϵ. Note that the ϵ-cardinality number is a nonincreasing function of ϵ. Clearly $m^{\text{non}}(R,\epsilon) \ge m^a(R,\epsilon)$. The 0-cardinality number is called the *cardinality number*.

Remark 2.1 As we mentioned in Corollary 3.1 of Chapter 3, if $r(R(N),N) = \epsilon$, then to find an ϵ-approximation we have to assume that the corresponding infima are attained. Similarly, if $r^{\text{non}}(R,m^{\text{non}}(\epsilon)) = \epsilon$ (or $r^a(R,m^a(\epsilon)) = \epsilon$), then we can find an ϵ-approximation only if the infimum in (2.8) is attained for some N and there exists an algorithm ϕ that uses N such that $e(\phi,N) = r(R(N),N)$, i.e., the infimum in (3.15) of Chapter 3 for $\rho = 0$ is also attained. ■

We illustrate these concepts by continuing our two examples.

Example: Binary Search (continued) Assume for simplicity that $m = 2^s$. Let N^a be an arbitrary adaptive information of cardinality n, that is,

$$N^a(f) = [Q(f,T_1),\dots, Q(f,T_n)], \qquad n \le s$$

for some adaptively chosen subsets $T_1, T_2,\dots, T_n$ of F. Note that $N^a(f)$ can take at most 2^n different values and that

$$\bigcup_{N^a(f)\in H} V(N^a,f) = F.$$

Thus there exists an element f such that $V(N^a,f)$ has at least 2^{s-n} elements. Further, if $n < s$, then there exists an element $g \in V(N^a,f)$ such that

$$|f-g| \ge 2^{s-n-1}.$$

This proves that

$$r(R(N^a), N^a) \ge r(\Phi(N^a), N^a) = r(N^a) \ge \begin{cases} 2^{-n-1} & \text{if} \quad n < s, \\ 0 & \text{if} \quad n = s. \end{cases} \tag{2.12}$$

We show that (2.12) is sharp, i.e., there exists information N such that for $n < s$, $r(N) = 2^{-n-1}$, and for $n = s$, $r(N) = 0$. Consider the adaptive information N^a defined by (2.5). Knowing $N^a(f)$, we can verify whether $f \in T_n$ or

$$f \in \{k_n + 2^{s-n} + 1, \ldots, k_n + 2 * 2^{s-n}\}.$$

In either case,

$$\mathrm{r}(N^a) = \begin{cases} 2^{-n-1} & \text{if} \quad n < s, \\ 0 & \text{if} \quad n = s. \end{cases}$$

This proves that N^a is nth optimal information (for $\Psi^a(n)$ and $\Phi(N^a)$), and the nth minimal radius is

$$r^a(n) = \begin{cases} 2^{-n-1} & \text{if} \quad n < s, \\ 0 & \text{if} \quad n = s. \end{cases}$$

From this we easily conclude that the ϵ-cardinality number is

$$m^a(\epsilon) = \min(s, \lceil \log \frac{1}{2\epsilon} \rceil).$$

(Here, as elsewhere in this book, log denotes logarithm to the base 2.)

Is adaptive information "stronger" than nonadaptive information for binary search? We prove that the answer is negative by exhibiting *nonadaptive information* N^{non} of cardinality n such that

$$r(N^{\text{non}}) = r^a(n).$$

Let

$$\mathrm{T}_i = \{g \in F\text{: the } (s - i) \text{ bit of } g - 1 = 0\}, \qquad i = 1, 2, \ldots, n.$$

We illustrate T_i for $s = 4$:

$F = T_0$	1	2	3	4	5	6	7	8	9	10	11	12	13	14	15	16
T_1	1	2	3	4	5	6	7	8								
T_2	1	2	3	4					9	10	11	12				
T_3	1	2			5	6			9	10			13	14		
T_4	1		3		5		7		9		11		13		15	

Define

$$N^{\text{non}}(f) = [Q(f,T_1),\ldots, Q(f,T_n)].$$

Knowing $N^{\text{non}}(f)$, we can determine the n most significant bits of f, which means that $V(N^{\text{non}},f)$ consists of 2^{s-n} elements. This yields

$$r(N^{\text{non}}) = r^a(n).$$

Thus the nonadaptive information N^{non} is nth optimal in the class of adaptive information. The binary search problem satisfies a condition which implies that nonadaption is as strong as adaption. This condition will be stated in Section 4.3.

We remark that in a parallel model of computation the time for computing $N^a(f)$ for the nth optimal adaptive information is proportional to n, whereas the time of computing $N^{\text{non}}(f)$ for the nth optimal nonadaptive information is constant, provided at least n processors are available. Thus, although $N^a(f)$ and $N^{\text{non}}(f)$ have the same local radii, their times in a parallel model are vastly different.

We stress that the proof of optimality for the binary search problem is rather straightforward. This is in a sharp contrast to other problems where determining the optimal adaptive information is very difficult or even open (see GTOA). ■

Example: Integration (continued) Let $N^a(f) = [f(x_1), f(x_2),\ldots, f(x_n)]$ be arbitrary adaptive information, $x_i = x_i(f(x_1),\ldots, f(x_{i-1}))$. Take $f = 0$ and let $z_i = x_i(0,\ldots, 0)$. Define nonadaptive information N^{non}:

$$N^{\text{non}}(f) = [f(z_1), f(z_2),\ldots, f(z_n)].$$

Then $N^a(0) = N^{\text{non}}(0)$ and therefore

$$r(N^a,0) = r(N^{\text{non}},0).$$

From Section 1.4 of Chapter 1 we know that

$$r(N^{\text{non}}) = r(N^{\text{non}},0).$$

We compute $r(N^{\text{non}},0)$. Without loss of generality, assume that z_i are ordered, $0 \le z_1 \le z_2 \le \ldots \le z_n \le 1$.

Define

$$f^+(x) = \begin{cases} -(x-z_1) & \text{if } 0 \le x \le z_1, \\ x-z_i & \text{if } z_i \le x \le (z_i + z_{i+1})/2, \\ -(x-z_{i+1}) & \text{if } (z_i + z_{i+1})/2 \le x \le z_{i+1}, \\ x-z_n & \text{if } z_n \le x \le 1. \end{cases}$$

Let $f^-(x) = -f^+(x)$. In Fig. 4.1 we illustrate the graphs of f^+ and f^-.

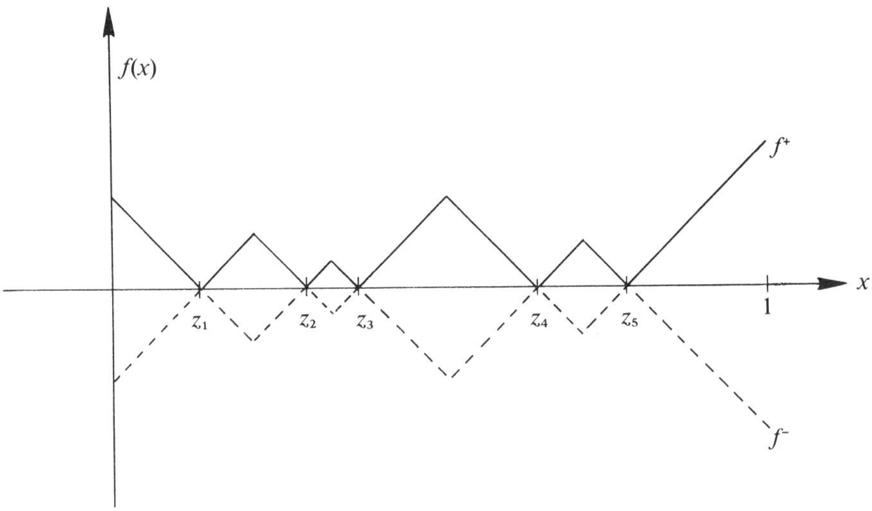

Figure 4.1

As in Section 1.4 of Chapter 1, it is clear that

$$f^-(x) \le \tilde{f}(x) \le f^+(x), \qquad \forall x \in [0,1], \qquad \forall \tilde{f} \in V(N^{\text{non}},0).$$

From this we conclude that

$$r(N^{\text{non}},0) = \int_0^1 f^+(x)\,dx = \tfrac{1}{2}[z_1^2 + (1 - z_n)^2] + \tfrac{1}{4}\sum_{i=1}^{n-1} (z_{i+1} - z_i)^2.$$

Let $h_i = z_{i+1} - z_i$, $i = 0, 1, \ldots, n$, with $z_0 = 0$ and $z_{n+1} = 1$. Then

$$\int_0^1 f^+(x)\,dx = g(\vec{h}) \overset{df}{=} \frac{1}{2}\,h_0^2 + \frac{1}{2}\,h_n^2$$

$$+ \frac{1}{4} \sum_{i=1}^{n-1} h_i^2 \geq \min_{h_0 + \ldots + h_n = 1} g(\vec{h}) = \frac{1}{4n},$$

where $\vec{h}$ denotes the vector $[h_0, \ldots, h_n]$. The minimum of g is uniquely attained for

$$z_{i+1}^* = \frac{2i+1}{2n}, \qquad i = 0, 1, \ldots, n-1.$$

Thus for any adaptive information N^a,

$$r(N^a) \geq r(N^a, 0) = r(N^{\text{non}}) \geq \frac{1}{4n}.$$

For the nonadaptive information N^{non},

$$N^{\text{non}}(f) = \left[f\left(\frac{1}{2n}\right), f\left(\frac{3}{2n}\right), \ldots, f\left(\frac{2n-1}{2n}\right) \right]$$

we have

$$r(N^{\text{non}}, 0) = r(N^{\text{non}}) = \frac{1}{4n}.$$

This proves that the nonadaptive information N^{non} is nth optimal in the class of *adaptive information*. Thus *nonadaption* is as strong as *adaption* for the integration problem. This property holds for a general class of linear problems, which will be discussed in Section 4.3 (see also GTOA). This example will be continued in Appendix A. ■

Although nonadaptive information is as strong as adaptive in both these examples, there are many problems where adaption is far stronger (see GTOA, Part A, Chapter 8, and Section 6.6 of Chapter 6).

4.3 ADAPTION VERSUS NONADAPTION

In Section 4.2 we showed that adaption does not help for binary search and integration. As we shall see, these two problems are examples of two classes of problems for which adaption is of the same strength as nonadaption, that is, $r^a(R,n) = r^{\text{non}}(R,n)$.

We begin with the first class of which the binary search problem is a member. For a given set H_1, let $\bar{\Psi}(H_1)$ denote the class of *all* operators U such that U: $F \to H_1$. Thus our class $\Psi(H_1)$ of primitive information operators is a subset, not necessarily proper, of $\bar{\Psi}(H_1)$.

For the binary search problem we have $H_1 = \{0,1\}$, and $\Psi(H_1)$ is the class of *all* single questions. This means that $\Psi(H_1) = \bar{\Psi}(H_1)$, and we shall prove that this is the property that makes adaption of the same strength as nonadaption.

THEOREM 3.1 Let $\Psi(H_1) = \bar{\Psi}(H_1)$, that is, $\Psi(H_1)$ is the class of *all* operators U, $U: F \to H_1$.

Then for any adaptive information N^a of cardinality n there exists a nonadaptive information operator N^{non} of cardinality n such that

$$N^a(f) = N^{\text{non}}(f), \qquad \forall f \in F. \tag{3.1}$$

Thus

$$r^a(R,n) = r^{\text{non}}(R,n) \qquad \text{and} \qquad m^a(R,\epsilon) = m^{\text{non}}(R,\epsilon), \tag{3.2}$$

which means that *adaption does not help.* ■

PROOF Let

$$N^a(f) = [U_1(f),\ U_2(f;y_1),\ldots,\ U_n(f;y_1,\ldots,\ y_{n-1})],$$

$y_i = U_i(f;y_1,\ldots,y_{i-1})$, be an adaptive information operator. We can rewrite $N^a(f)$ as

$$N^a(f) = [U_1(f),\ U_2(f;U_1(f)),\ldots,\ U_n(f;U_1(f),\ldots,U_{n-1}(f;\ldots)].$$

Define

$$W_1(f) = U_1(f),$$

$$W_2(f) = U_2(f;U_1(f)),$$

$$\vdots$$

$$W_n(f) = U_n(f;U_1(f),\ldots,\ U_{n-1}(f;\ldots)).$$

Then $W_i : F \to H_1$ and $W_i \in \bar{\Psi}(H_1)$. Since every mapping is primitive, then $W_i \in \Psi(H_1)$. Define the *nonadaptive information operator* N^{non} as

$$N^{\text{non}}(f) = [W_1(f),\ W_2(f),\ldots,\ W_n(f)].$$

Then $N^{\text{non}} \in \Psi^{\text{non}}(n)$ and, of course,

$$N^a(f) = N^{\text{non}}(f), \qquad \forall f \in F,$$

which proves (3.1). Since (3.2) follows immediately from (3.1), the proof is completed. ■

Theorem 3.1 states that there is no difference between adaptive and nonadaptive information operators as long as $\Psi(H_1) = \bar{\Psi}(H_1)$. Thus adaption can be more powerful than nonadaption only if $\Psi(H_1) \neq \bar{\Psi}(H_1)$.

We comment on the assumption $\Psi(H_1) = \bar{\Psi}(H_1)$. In many cases it is unrealistic to assume that every operator is a primitive information operator. Furthermore, the "cost" of primitive information operators may vary (see Chapter 5) and the cost of adaptive information N^a can be much smaller than the cost of the corresponding nonadaptive information N^{non} in (3.1). For example, the cost of computing $U_i(f; y_1,\dots, y_{i-1})$ may be significantly smaller than the cost of computing $W_i(f)$ defined in the proof of Theorem 3.1.

Remark 3.1 We show that the primitive information operators W_i satisfy the following property. Let $(z_1,z_2,\dots,z_i) \in H_1^i$. Define $U_i^{-1}(z_1,z_2,\dots,z_i) = \{f \in F: U_1(f) = z_1,\ U_2(f;z_1) = z_2,\dots,\ U_i(f;z_1,\dots,z_{i-1}) = z_i\}$, for $i = 1,2,\dots,n$. Although U_i^{-1} is not precisely an inverse of U_i, we choose this notation as suggestive. Observe that $U_i^{-1}(z_1,\dots,z_i)$ can be empty. The sets $U_i^{-1}(z_1,\dots,z_i)$ are disjoint, that is, $(z_1^1,\dots,z_i^1) \neq (z_1^2,\dots,z_i^2)$ implies $U_i^{-1}(z_1^1,\dots,z_i^1) \cap U_i^{-1}(z_1^2,\dots,z_i^2) = \emptyset$, and

$$\bigcup_{(z_1,\dots,z_i)\in H_1^i} U_i^{-1}(z_1,\dots,z_i) = F, \qquad \forall i \in [1,n].$$

Let $f \in F$. Then there exists a unique $(z_1,z_2,\dots,z_i)$ such that $f \in U_i^{-1}(z_1,z_2,\dots,z_i)$. We show that

$$W_i(f) = z_i, \qquad \forall i \in [1,n].$$

Indeed, let $i = 1$. Then $f \in U_1^{-1}(z_1)$ iff $U_1(f) = z_1$. Since $W_1(f) = U_1(f)$, then $W_1(f) = z_1$. Assume now inductively that $W_j(f) = z_j$ for $j = 1,2,\dots,i-1$. Then $f \in U_i^{-1}(z_1,z_2,\dots,z_i)$ implies that $U_i(f; z_1,\dots,z_{i-1}) = z_i$ where

$$z_j = U_j(f; U_1(f),\dots, U_{j-1}(f;\dots)), \qquad j = 1,\dots,i-1.$$

From the definition of W_i we have $W_i(f) = U_i(f,z_1,\dots,z_{i-1}) = z_i$, which completes the proof. ■

We turn to a second class of problems for which adaption also does not help. The integration problem belongs to this class. Recall that for the integration problem, $\Psi(H_1)$ was defined as the class of all single function evaluations, that is, $H_1 = \mathbb{R}$ and $\Psi(H_1)$ is the class of *linear* functionals of the special form $U(f) = f(x)$ for some $x = x(U) \in [0,1]$. Of course, in this case $\Psi(H_1) \neq \bar{\Psi}(H_1)$ (since $\bar{\Psi}(H_1)$ contains, for example, nonlinear functionals). Thus Theorem 3.1 is not applicable, although adaption still does not help. Why? Consider the proof for the integration problem. For any adaptive information N^a we constructed nonadaptive information N^{non} by taking fixed numbers $y_i = 0$ and proving that for $f = 0$ we have $r(N^a,f) = r(N^{\text{non}},f) = r(N^{\text{non}})$.

We now generalize this construction. For simplicity we assume that $R(N) = \Phi(N)$, $\forall N$. Let

$$N^a(f) = [U_1(f),\ U_2(f;y_1),\ \ldots,\ U_n(f;\ y_1,\ \ldots,\ y_{n-1})], \tag{3.3}$$

where $y_i = U_i(f;y_1,\ \ldots,y_{i-1})$ be an adaptive information operator. For a fixed f^* from F let

$$y_i^* = U_i(f^*;y_1^*,\ldots,y_{i-1}^*), \qquad i = 1,\ 2,\ldots,\ n-1.$$

Define the nonadaptive information operator as

$$N_{f^*}^{\text{non}}(f) = [U_1(f),\ U_2(f;y_1^*),\ldots,\ U_n(f;y_1^*,\ldots,\ y_{n-1}^*)]. \tag{3.4}$$

That is, $N_{f^*}^{\text{non}}$ is derived from N^a by taking the fixed y_i^*which is generated by f^*. The cardinality and the "structure" of $N_{f^*}^{\text{non}}$ is the same as N^a. When is $N_{f^*}^{\text{non}}$ of the same strength as N^a, i.e., when is the radius of $N_{f^*}^{\text{non}}$ the same or comparable to the radius of N^a? To answer this question we define the concept of c-hard. Namely we say that f is *c-hard for N* iff

$$r(N,f) \geq c\ r(N), \qquad c > 0. \tag{3.5}$$

Note that $c \leq 1$, and if c is close to 1, then the radius $r(N,f)$ for f is close to the radius of information. For $c = 1$, (3.5) implies that the supremum in (3.4) of Chapter 1 is attained for f.

THEOREM 3.2 Let N^a and $N_{f^*}^{\text{non}}$ be defined by (3.3) and (3.4). If f^* is c-hard for $N_{f^*}^{\text{non}}$, then

$$r(N_{f^*}^{\text{non}}) \leq \tfrac{1}{c}\, r(N^a). \tag{3.6}$$

■

PROOF Observe that $N^a(f^*) = N_{f^*}^{\text{non}}(f^*)$, which implies that $V(N^a,f^*) = V(N_{f^*}^{\text{non}},f^*)$ and $r(N^a,f^*) = r(N_{f^*}^{\text{non}},f^*)$. From (3.5) we therefore have

$$r(N_{f^*}^{\text{non}}) \leq \tfrac{1}{c}\ r(N_{f^*}^{\text{non}},f^*) = \tfrac{1}{c}\ r(N^a,f^*) \leq \tfrac{1}{c}\ r(N^a),$$

which proves (3.6). ■

The essence of Theorem 3.2 is that to find c we need to analyze only the nonadaptive information $N_{f^*}^{\text{non}}$, which is simpler than analyzing the significantly more complicated structure of adaptive information. Furthermore, if c does not depend on $N_{f^*}^{\text{non}}$ (or ideally if $c = 1$ for any $N_{f^*}^{\text{non}}$), we conclude that adaption does not help. More precisely we have

COROLLARY 3.1 Suppose there exists $c \in [0,1]$ such that for any $N^a \in \Psi^a(n)$ one can find f^* which is c-hard for $N_{f^*}^{\text{non}}$. Then

$$\begin{aligned} r^a(n) &\leq r^{\text{non}}(n) \leq \tfrac{1}{c}\, r^a(n), \\ m^a(\epsilon) &\leq m^{\text{non}}(\epsilon) \leq m^a(c\,\epsilon). \end{aligned} \tag{3.7}$$

■

PROOF The left-hand side of (3.7) is always true. Taking the infimum in (3.6) over $N^a \in \Psi^a(n)$, we get

$$r^{\text{non}}(n) \leq \inf_{f^*} r(N_{f^*}^{\text{non}}) \leq \tfrac{1}{c}\, r^a(n).$$

From this and the definition of $m^a(\epsilon)$ and $m^{\text{non}}(\epsilon)$, (3.7) easily follows. ■

Note that for $c = 1$ we have equality in (3.7). In the next example $c = 1/2$.

Example: Normed Linear Space (continued) Let F be a balanced and convex subset of a linear space F_1, that is, $f \in F$ implies $-f \in F$ and $f_1, f_2 \in F$ implies $tf_1 + (1-t)f_2 \in F$, $\forall t \in [0,1]$. Let G be a normed linear space and $S : F \to G$ be a linear operator, that is, $S(t_1 f_1 + t_2 f_2) = t_1 S(f_1) + t_2 S(f_2)$, $\forall f_1, f_2 \in F_1$ and scalars t_1, t_2. As in Section 1.3 of Chapter 1, define

$$S(f,\epsilon) = \begin{cases} \{g \in G : \| S(f) - g \| < \epsilon\} & \text{for} \quad \epsilon > 0, \\ \{S(f)\} & \text{for} \quad \epsilon = 0. \end{cases}$$

Let $H_1 = \mathbb{C}$ be the complex field and let $\Psi(H_1)$ be the class of *linear* functionals U. Then any adaptive information N^a from $\Psi^a(n)$ is of the form

$$N^a(f) = [U_1(f),\ U_2(f;y_1), \ldots,\ U_n(f;y_1,y_2,\ldots,y_{n-1})],$$

where $y_i = U_i(f;y_1,\ldots,y_{i-1})$ and $U_i(\cdot\,;y_1,\ldots,y_{i-1})$ is a linear functional. Let $f^* = 0$. Due to the linearity of $U_i(\cdot\,;y_1,\ldots,y_{i-1})$ we have $y_i^* = 0$ and (3.4) becomes

$$N_0^{\text{non}}(f) = [U_1(f),\ U_2(f;0), \ldots, U_n(f;0,\ldots,0)]. \tag{3.8}$$

Note that N_0^{non} is a *linear operator*. Let $f_1, f_2 \in V(N_0^{\text{non}}, f)$. Then $h = f_1 - f_2 \in \ker N_0^{\text{non}}$ and $h/2 \in F$. This yields

$$\operatorname{diam}(S(V_0^{\text{non}}, f)) \leq 2 \sup_{h \in \ker N_0^{\text{non}} \cap F} \| Sh \| = \operatorname{diam}(S(V(N_0^{\text{non}}, 0))). \tag{3.9}$$

From (3.9) and Lemma 3.2 of Chapter 3 we conclude

$$r(N_0^{\text{non}}) \le \operatorname{diam}(S(V(N^{\text{non}},0))) \le 2r(N_0^{\text{non}},0).$$

This corresponds to (3.5) with $c = 1/2$. Thus we have proven

THEOREM 3.3 For the normed linear space setting, adaption is at most twice more effective than nonadaption, i.e.,

$$r^{\text{non}}(n) \le 2r^{a}(n). \qquad \blacksquare$$

This theorem was established by Bakhvalov [71] who assumed that S is a linear functional, by Gal and Micchelli [80] who used a different technique, and by the authors of GTOA who also showed numerous examples for which $r^{\text{non}}(n) = r^a(n)$. One of the examples is the integration problem studied throughout this book. This example will be continued in Section 4. ■

4.4 OPTIMAL APPROXIMATE INFORMATION

In this section we generalize the concept of optimal information to approximate information. Assume that E is the information error operator as defined in Chapter 2. Let N be the exact information operator of the form (2.3). Let N_ρ be an approximate information operator, i.e., $N_\rho : F \to H$, and $N_\rho(f) \in E(N(f),\rho)$, $\forall f \in H$.

Assume that $R(N_{\rho,E})$ is a class of realizable algorithms ϕ that use N_ρ. Recall that the radius $r(R(N_{\rho,E}), N_{\rho,E})$ is defined by (3.19) of Chapter 3.

We define the *nth minimal radius of approximate information* (for $\Psi^{\text{non}}(n)$, R, and ρ) or ($\Psi^a(n)$, R, and ρ) as

$$r^{\text{non}}(R,n,\rho) = \inf_{N \in \Psi^{\text{non}}(n)} r(R(N_{\rho,E}), N_{\rho,E})$$

$$(\text{or } r^a(R,n,\rho) = \inf_{N \in \Psi^{a}(n)} r(R(N_{\rho,E}), N_{\rho,E})). \tag{4.1}$$

As before, the symbol R does not denote the class of realizable algorithms, but is to emphasize that we are using classes of realizable algorithms.

An information operator N is *nth optimal approximate information* (for $\Psi^{\text{non}}(n)$, R, and ρ) or ($\Psi^a(n)$, R, and ρ) iff

$$N \in \Psi^{\text{non}}(n) \qquad (\text{or } N \in \Psi^a(n)) \tag{4.2}$$

and

$$r(R(N_{\rho,E}),N_{\rho,E}) = r^{\text{non}}(R,n,\rho) \qquad (\text{or } r(R(N_{\rho,E}),N_{\rho,E}) = r^a(R,n,\rho)). \tag{4.3}$$

For $\rho = 0$, the above definitions coincide with (2.8), (2.9), and (2.10) since $N_{0,E} = N$. Note that we have defined nth optimal approximate information to be the *exact* information which minimizes the radii of approximate information operators of a given cardinality. Similarly to (2.11), we define the ϵ-cardinality number (for approximate nonadaptive or adaptive information, and for R and ρ) as

$$m^{\text{non}}(R,\rho,\epsilon) = \min\{n: r^{\text{non}}(R,\rho,n) \le \epsilon\}$$
$$(\text{or } m^{a}(R,\rho,\epsilon) = \min\{n: r^{a}(R,\rho,n) \le \epsilon\}). \tag{4.4}$$

If $R(N_{\rho,E})$ is the class of idealized algorithms for any N, then we do not list R in (4.1) through (4.4). As in Section 4.2, the 0-cardinality number is simply called the cardinality number. We emphasize that all these concepts depend on the choice of primitive information operators. We comment on the nth minimal radii of approximate information for adaptive and nonadaptive classes of information. From Theorem 3.1 it follows that for any N^a we can find N^{non} such that they are equal, provided $\Psi(H_1) = \bar{\Psi}(H_1)$. This implies that $\Psi^{\text{non}}(n) = \Psi^{a}(n)$, and since the corresponding sets $E(N^a(f),\rho)$ and $E(N^{\text{non}}(f),\rho)$ are also equal, we immediately have

COROLLARY 4.1 If $\Psi(H_1) = \bar{\Psi}(H_1)$, then

$$r^{\text{non}}(R,\rho,n) = r^{a}(R,\rho,n), \qquad m^{\text{non}}(R,\rho,\epsilon) = m^{a}(R,\rho,\epsilon). \qquad \blacksquare$$

We now generalize Theorem 3.2 for approximate information. Similarly to (3.5), we say that f is *c-hard for N, E,* and ρ iff

$$r(N_{\rho,E},f) \ge c\, r(N_{\rho,E}). \tag{4.5}$$

THEOREM 4.1 Let N^a and $N^{\text{non}}_{f^*}$ be defined by (3.3) and (3.4). If f^* is c-hard for $N^{\text{non}}_{f^*}$,E, and ρ, then

$$r(N^{\text{non}}_{f^*\rho,E}) \le \tfrac{1}{c}\, r(N^{a}_{\rho,E}). \tag{4.6}$$

PROOF Since $N^a(f^*) = N^{\text{non}}_{f^*}(f^*)$, then $E(N^a(f^*),\rho) = E(N^{\text{non}}_{f^*}(f^*),\rho)$ and

$$r(N^{\text{non}}_{f^*\rho,E}) \le \tfrac{1}{c}\, r(N^{\text{non}}_{f^*\rho,E},f^*) = \tfrac{1}{c}\, r(N^{a}_{\rho,E},f^*) \le \tfrac{1}{c}\, r(N^{a}_{\rho,E}).$$

This proves (4.6). $\blacksquare$

From Theorem 4.1 we obtain

COROLLARY 4.2 Suppose there exists $c \in (0,1]$ such that for any $N^a \in \Psi^a(n)$ one can find f^* which is c-hard for $N^{\text{non}}_{f^*}$, E, and ρ.

Then

$$r^a(\rho,n) \le r^{\text{non}}(\rho,n) \le \tfrac{1}{c}\, r^a(\rho,n),$$
$$m^a(\rho,\epsilon) \le m^{\text{non}}(\rho,\epsilon) \le m^a(\rho,c\epsilon). \tag{4.7}$$

■

The proof is essentially the same as the proof of Corollary 3.1 and is therefore omitted.

We illustrate Theorem 4.1 and Corollary 4.2 by an example with $c \ge 1/2$.

Example: Normed Linear Space (continued) Let H be a normed linear space. Define

$$E(h,\rho) = \{g \in H: \| h - g \| \le \rho\}$$

as the ball of center h and radius ρ. Let $N_\rho(f)$ be approximate information. Thus, $\| N_\rho(f) - N(f) \| \le \rho$. As in Section 4.3, assume that F is balanced and convex, S is a linear operator, and $\Psi(H_1)$ consists of linear functionals. Let N be a linear information operator. Micchelli and Rivlin [77] proved that

$$r(N_{\rho,E}) = c^{-1} \sup\{\| Sh \| : h \in F \text{ and } \| Nh \| \le \rho\}, \tag{4.8}$$

where $c \in [1/2,1]$. It is easy to check that for $f = 0$, we have

$$r(N_{\rho,E},0) \ge c\, r(N_{\rho,E}). \tag{4.9}$$

Consider the adaptive information N^a defined by (3.3) and the nonadaptive information N^{non} defined by (3.4) with $f^* = 0$. Since N^{non} is linear, (4.8) and (4.9) yield that f^* is c-hard for N, E, and ρ with c equal to at least 1/2. Thus adaptive approximate information is at most twice more effective than nonadaptive approximate information for the normed linear space setting. This example will be continued in Appendix A. ■

CHAPTER FIVE

COMPLEXITY

5.1 INTRODUCTION

The first four chapters are independent of a model of computation and therefore of complexity. As we remarked earlier, this independence makes many of the results even stronger. In this chapter we define our model of computation and introduce our complexity concepts. The basic concept is problem complexity, which is the minimal cost of obtaining an ϵ-approximation.

As part of the model of computation, we specify our primitives. A *permissible algorithm* is one that uses a finite number of primitive operations. In Section 5.5, the relations among the major classes of algorithms (idealized, realizable, permissible) are examined.

We comment on the relation of permissible algorithm to the notion of algorithm in certain branches of theoretical computer science. We use Turing machines for illustration. The model of computation is the Turing machine model. The class of permissible algorithms consists of algorithms which can be executed with finite complexity on a Turing machine.

We summarize the results of this chapter. Our model of computation is given in Section 5.2. It is essentially the same as in GTOA and is repeated here for the reader's convenience. The model is based on the concepts of primitive operation, permissible information, and permissible algorithm. Complexity is defined in Section 5.3. The basic concepts are problem complexity and optimal complexity algorithm. The first two subsections of Section 5.4 demonstrate how complexity depends on the model of computation and how it may be decreased by precomputation. In the third subsection we exhibit relations between complexity concepts and ϵ-entropy for discrete information operators.

The concluding section discusses the relations among the classes of idealized, realizable, and permissible algorithms.

5.2 MODEL OF COMPUTATION

We briefly outline our model of computation, which consists of a set of primitive operations, permissible information, and permissible algorithms.

(i) Let p be a *primitive operation*. Examples of primitive operations include arithmetic operations, comparisons, taking the maximum of n numbers, the evaluation of a radical, an integral, a linear functional or a nonlinear functional. Let comp(p) be the complexity of p. We assume that comp(p) is finite. Suppose that P is a given collection of primitives. The choice of P and comp(p), $p \in P$, is arbitrary and depends on the particular problem being solved.

(ii) Let N_ρ be an approximate information operator. We say that N_ρ is *permissible* with respect to P if there exists a program that uses a finite number of primitive operations from P to compute $N_\rho(f)$ for all $f \in F$. If $N_\rho(f)$ requires the evaluation of primitives $p_1, p_2, \ldots, p_k$, then

$$\text{comp}(N_\rho(f)) = \sum_{i=1}^{k} \text{comp}(p_i).$$

We call comp($N_\rho(f)$) the *information complexity* of computing $N_\rho(f)$.

(iii) Let ϕ be an algorithm that uses the permissible information N_ρ. To evaluate $\phi(N_\rho(f))$ we

(a) compute $y = N_\rho(f)$;

(b) compute $\phi(y)$.

The complexity of computing y is given by (ii). We say that ϕ is *permissible* with respect to P if there exists a program that uses a finite number of primitive operations from P to compute $\phi(y)$ for all $y = N_\rho(f)$, $f \in F$. Let comp($\phi(y)$) be the *combinatory complexity* of computing $\phi(y)$. Thus, if $\phi(y)$ requires the evaluation of primitives $q_1, q_2, \ldots, q_j$, then

$$\text{comp}(\phi(y)) = \sum_{i=1}^{j} \text{comp}(q_i).$$

Remark 2.1 Note that an idealized algorithm or a realizable algorithm need not be permissible. The class of idealized algorithms includes the classes of realizable and permissible algorithms. ■

Remark 2.2 The set P of primitive operations is often defined in such a way that it consists of the four arithmetic operations on real numbers, each with unit complexity. That is, we assume that we can *exactly* add, subtract, multiply, or divide two real numbers with unit cost. This is the real number model of computation. Although this model does not exist in a finite universe, it is a very useful mathematical abstraction which is widely used in algebraic complexity, e.g., in the famous matrix multiplication problem.

Of course, fixed-precision floating-point arithmetic, rather than real arithmetic, is almost universally used in practice. Although algorithm stability is a very important issue and only approximate results can be computed by using fixed-precision floating-point arithmetic, complexity results are essentially the same as in the real number model. Therefore we generally use the real number model rather than the fixed-precision floating-point model to avoid round-off issues (for further discussion, see Traub and Woźniakowski [82]). ■

5.3 COMPLEXITY CONCEPTS

Let N_ρ be a permissible information operator and let $R(N_\rho)$ be a class of realizable algorithms such that $r(R(N_\rho),N_\rho) \le \epsilon$. Corollary 3.1 of Chapter 3 states that we can find an ϵ-approximation by using N_ρ (assuming that when $r(R(N_\rho),N_\rho) = \epsilon$, condition (ii) of this corollary holds). Let $R(N_\rho,\epsilon)$ be the subclass of the class $R(N_\rho)$ of realizable algorithms ϕ, which are permissible and whose errors $e(\phi,N_\rho)$ do not exceed ϵ. Note that

$$r(R(N_\rho),N_\rho) > \epsilon \quad \text{implies} \quad R(N_\rho,\epsilon) = \emptyset.$$

We define the *complexity of an algorithm* ϕ as

$$\text{comp}(\phi) = \sup_{f \in F} (\text{comp}(N_\rho(f)) + \text{comp}(\phi(N_\rho(f)))). \tag{3.1}$$

We say that $\text{comp}(R,N_\rho,\epsilon)$ is the *ϵ-complexity* (for $R(N_\rho)$ and N_ρ) iff

$$\text{comp}(R(N_\rho),N_\rho,\epsilon) = \inf\{\text{comp}(\phi) : \phi \in R(N_\rho,\epsilon)\} \tag{3.2}$$

with the convention $\inf \emptyset = +\infty$.

We say that ϕ^{oc}, $\phi^{oc} \in R(N_\rho,\epsilon)$ is an *optimal complexity algorithm* (for $R(N_\rho)$ and N_ρ) iff

$$\text{comp}(\phi^{oc}) = \text{comp}(R(N_\rho),N_\rho,\epsilon). \tag{3.3}$$

Let Ψ be a class of permissible approximate information operators. We say that $\text{comp}(R,\Psi,\epsilon)$ is the *ϵ-complexity in the class* Ψ (for R) iff

$$\text{comp}(R,\Psi,\epsilon) = \inf_{N_\rho \in \Psi} \text{comp}(R(N_\rho),N_\rho,\epsilon). \tag{3.4}$$

We shall refer to the ϵ-complexity in the class Ψ as the *problem complexity.*

We say that ϕ^{oc} is an *optimal complexity algorithm* for the class Ψ (for R) iff

$$\text{comp}(\phi^{oc}) = \text{comp}(R,\Psi,\epsilon), \tag{3.5}$$

and ϕ^{oc} uses an approximate information operator N_ρ from Ψ and belongs to $R(N_\rho)$. As always, if $R(N_\rho) = \Phi(N_\rho)$, we do not list $R(N_\rho)$ in the above formulas. If $\epsilon = 0$, then 0-complexity is simply called the complexity and we also do not list ϵ. Thus, for instance, comp(N_ρ) denotes comp$(\Phi(N_\rho),N_\rho,0)$.

5.4 REMARKS AND EXAMPLES

This section contains three subsections with the following material. In subsection (i) we deal with the division problem to illustrate how ϵ-complexity depends on the set of primitive operations. We prove that the ϵ-complexity is very close to log log$(1/\epsilon)$ if three arithmetic operations (addition, subtraction, and multiplication) are allowed. In subsection (ii) we exhibit the relation between ϵ-complexity and the ϵ-cardinality number under certain assumptions. In subsection (iii) we show how the ϵ-cardinality number and ϵ-complexity are related to ϵ-entropy for all discrete information operators.

(i) We stress that all the complexity concepts depend on the set P of primitive operations. It may happen that for one set P_1 of primitive operations the ϵ-complexity is large, whereas for a different set P_2 it is small. We illustrate this by the example of division.

Example 4.1 Let $F = [-1,1]$ and $G = \mathbb{R}$. Define

$$S(f) = \frac{1}{3 - f}$$

and $S(f,\epsilon) = \{g \in G: | S(f) - g | \le \epsilon\}$. Let $N(f) = f$ and $\rho = 0$. Obviously, $r(N) = 0$. Consider the set of primitive operations

$$P = P_1 = \{+, -, *\}$$

and let the complexity of any primitive be equal to unity. Thus, we can perform only three arithmetic operations, and an algorithm ϕ is permissible iff $\phi(f)$ is a polynomial. Therefore we define the class $R_1 = R_1(N)$ of realizable algorithms as the class of polynomials, i.e., $R_1 = \{w\colon w$ is a polynomial$\}$. Note that $r(R_1,N) = 0$ since $S(f)$ is a continuous function of f and can be uniformly approximated by polynomials. The class $R_1(N,\epsilon)$ of permissible algorithms is a subset of R_1 and consists of polynomials that approximate $S(f)$ with error no larger than ϵ. Let ϕ be a permissible algorithm, i.e., let ϕ be a polynomial of degree, say, n. Let Π_n be the set of all polynomials of degree at most n. Then

$$e(\phi,N) = \sup_{f\in F} | S(f) - \phi(f) | \ge E_n(f) \stackrel{df}{=} \inf_{w\in\Pi_n} \sup_{f\in F} | S(f) - w(f) | .$$

It is known (see Achieser [56, pp. 58–60] that

$$E_n(f) = \frac{(3 - \sqrt{8})^n}{8}.$$

The minimal n for which $E_n(f) \leq \epsilon$ is given by

$$n^* = \left\lceil \frac{\log[1/(8\epsilon)]}{\log[1/(3 - \sqrt{8})]} \right\rceil \simeq 0.4 \log(1/\epsilon)$$

Thus, if $e(\phi,N) \leq \epsilon$ and ϕ is a polynomial of degree n, then $n \geq n^*$. Note that to compute $\phi(f)$ we have to perform at least $\lceil \log n \rceil$ multiplications (see Borodin and Munro [75, Theorem 2.4.7]). This implies that the complexity of any algorithm ϕ from $R_1(N,\epsilon)$ satisfies

$$\text{comp}(\phi) \geq \lceil \log n^* \rceil = \log \log 1/\epsilon(1 + o(1)). \tag{4.1}$$

Does there exist a permissible algorithm whose complexity is close to log log $(1/\epsilon)$? The answer is YES and is provided by an algorithm that is based on Newton iteration applied to the nonlinear equation $(1/x) - (3 - f) = 0$. That is, we define the sequence $\{x_i\}$ as

$$x_0 = 1/3, \qquad x_{i+1} = x_i(2 - x_i(3 - f)), \qquad i = 0, 1, \ldots. \tag{4.2}$$

It is easy to check that

$$x_{i+1} - \frac{1}{3 - f} = (f - 3)\left(x_i - \frac{1}{3 - f}\right)^2$$

$$= [(f - 3)x_0 + 1]^{2^{i+1}-1}\left(x_0 - \frac{1}{3 - f}\right).$$

Since $|(f - 3)x_0 + 1| \leq 1/3$, we have

$$\left| x_i - \frac{1}{3 - f} \right| \leq \frac{3^{1-2^i}}{6}.$$

Thus $|x_i - 1/(3 - f)| \leq \epsilon$ whenever

$$i \geq k = \lceil \log(\log (1/2\epsilon) / \log 3) \rceil = \log \log \tfrac{1}{\epsilon}(1 + o(1)).$$

Define

$$\phi(f) = x_k. \tag{4.3}$$

Then ϕ is permissible and $e(\phi,N) \leq \epsilon$. The complexity of ϕ is equal to the cost of performing k steps of (4.2). At each step we perform one subtraction and two multiplications since $(3 - f)$ need be computed only once. Thus

$$\text{comp}(\phi) = 3k + 1 = 3 \log \log \tfrac{1}{\epsilon} (1 + o(1)). \tag{4.4}$$

From (4.1) and (4.4) we conclude that the ϵ-complexity satisfies

$$\text{comp}(R_1,N,\epsilon) = \Theta(\log \log \tfrac{1}{\epsilon}) \tag{4.5}$$

and algorithm (4.3) is a nearly optimal complexity algorithm. Thus, the ϵ-complexity of the division problem is essentially $\log \log(1/\epsilon)$ if $+, -, *$ are primitives. Note that the ϵ-complexity tends (very slowly) to infinity as ϵ goes to zero.

Consider now the second set of primitive operations

$$P = P_2 = \{+, -, *, /\}$$

and let the complexity of any primitive be equal to unity. Let $R_2 = R_2(N)$ be defined by $R_2 =$ the set of rational functions. Then the algorithm

$$\phi(f) = S(f) = \frac{1}{3 - f}$$

is permissible and its complexity is two. It is easy to see that ϕ is an optimal complexity algorithm and

$$\text{comp}(R_2,N,\epsilon) = 2, \qquad \forall \epsilon \geq 0. \tag{4.6}$$

This shows that the ϵ-complexity depends essentially on the set of primitive operations. ■

(ii) We discuss relations among several major complexity and information concepts. Assume for simplicity that $\rho = 0$. Let

$$N(f) = [U_1(f), U_2(f), \ldots, U_n(f)],$$

where the U_i are permissible primitive information operators. If $\text{comp}(U_i) = c_1$, $i = 1, 2, \ldots, n$, for some c_1, then

$$\text{comp}(N) = nc_1, \tag{4.7}$$

which shows how the information complexity depends on the cardinality of N. There are many problems for which the assumption $\text{comp}(U_i) \equiv \text{const}$ is realistic (see GTOA).

Let ϕ be an algorithm that uses N. The combinatory complexity $\phi(N(f))$ may sometimes be lowered by precomputation. We explain this by an example.

Example 4.2 Let $U_i(f)$ be a complex number and let ϕ be of the form

$$\phi(N(f)) = \sum_{i=1}^{n} U_i(f)\, g_i$$

for some elements g_i from G. Such an algorithm is said to be *linear*. Numerous examples of linear algorithms, which are also optimal error algorithms, may be found in GTOA. The elements g_i are independent of f but may depend on N and S. They can be precomputed, and the computation of $\phi(N(f))$ then requires only n multiplications and $n - 1$ additions. If we assume that multiplication and addition are primitives with unit complexity, then the combinatory complexity $\text{comp}(\phi(N(f)))$ is at most $2n - 1$. Usually c_1 in (4.7) is much larger than unity, yielding

$$\text{comp}(\phi(N(f))) << \text{comp}(N(f)). \qquad ■$$

Example 4.2 shows that we can sometimes assume that $\text{comp}(\phi(N(f))) <<$ $\text{comp}(N(f))$. Then

$$\text{comp}(\phi) \simeq \text{comp}(N,\epsilon) \simeq nc_1. \tag{4.8}$$

Thus ϕ is a *nearly optimal complexity algorithm* and the ϵ-complexity is a linear function of the cardinality of N.

Let Ψ be the class of adaptive information operators of finite cardinality. If (4.8) holds for every N, then the ϵ-complexity in the class Ψ is closely related to the ϵ-cardinality number $\text{m}^a(\epsilon)$ since

$$\text{comp}(\Psi,\epsilon) \simeq m^a(\epsilon)\, c_1. \tag{4.9}$$

The algorithm ϕ that satisfies (4.8) for $m^a(\epsilon)$th optimal information, is a nearly optimal complexity algorithm for the class Ψ. Many practical examples for which (4.8) and (4.9) hold may be found in GTOA.

We stress that the relation between the ϵ-complexity and the ϵ-cardinality number holds under the *equal information complexity* assumption (4.7). For many problems this assumption is false (see, for example, Traub and Woźniakowski [80c, Section 6] and a problem studied by Micchelli and Miranker [75] which is discussed in GTOA, Part A, pp. 170–173). If the equal information complexity assumption is false, then minimal cardinality of information does not guarantee minimal or close to minimal complexity.

(iii) There are important relations between complexity concepts and ϵ-entropy for the normed linear space setting (see GTOA). In the following p-ary search example we exhibit similar relations for discrete information operators.

Example 4.3 Let S, F, G be defined analogously to their definitions in the Binary Search Example. Let H_1 be a set consisting of p elements, $p > 1$, and assume that every U, $U: F \to H_1$, is permissible. Assume additionally that

$$\text{comp}(U) = c_1, \qquad \forall U. \tag{4.10}$$

Then for every permissible N, $N = [U_1, \ldots, U_n]$, we have

$$\text{comp}(N) = c_1 n.$$

Remark 4.1 Note that $N(f)$ is a vector of n components and each component has at most p different values from H_1. Thus $N(f)$ can be represented by $n \lceil \log p \rceil$ bits and c_1 can be taken as

$$c_1 = \lceil \log p \rceil .$$

■

What is the ϵ-complexity of problem S? To answer this question it is useful to define ϵ-entropy. We proceed as follows. Let $f \in F$ and $g \in G$. Let

$$\text{dist}_G(f,g) = \inf\{\delta: g \in S(f,\delta)\}$$

and

$$J(g,\delta) = \{f \in F: \text{dist}_G(f,g) \le \delta\} \qquad (= \{f \in F: g \in S(f,\delta + \eta), \forall \eta > 0\}).$$

Then $J(g,\delta)$ can be treated as a "ball" with center g and radius δ.

Let $X \subset F$. Then $H(\epsilon,X)$ is the *ϵ-entropy of X* iff

$$H(\epsilon,X) = \log_2 k, \tag{4.11}$$

where

$$k = \min\{j: \exists g_1, \ldots, g_j \in G,\ X \subset \bigcup_{i=1}^{j} J(g_i,\epsilon)\}.$$

Thus k is the minimal number of balls with radii not greater than ϵ which cover the set X. By H^{-1} we mean

$$H^{-1}(k,X) = \inf\{\epsilon: \exists g_1, \ldots, g_k \in G,\ X \subset \bigcup_{i=1}^{k} J(g_i,\epsilon)\}. \tag{4.12}$$

THEOREM 4.1

$$r^{\text{non}}(n) = r^a(n) = H^{-1}(p^n,F); \tag{4.13}$$

$$m^{\text{non}}(\epsilon) = m^a(\epsilon) = \left\lceil \frac{H(\epsilon,F)}{\log p} \right\rceil . \tag{4.14}$$

■

PROOF From (3.2) of Chapter 4 we know that $r^{\text{non}}(n) = r^a(n)$ and $m^{\text{non}}(\epsilon) = m^a(\epsilon)$.

We first prove that $r^a(n) \geq H^{-1}(p^n,F)$. Let $N = [U_1,\ldots, U_n]$ and assume without loss of generality that $r(N)$ is finite and take $b > r(N)$. Then there exists $\phi \in \Phi(N)$ such that $e(\phi,N) < b$. Let

$$\{g_1,\ldots, g_k\} = \bigcup_{f \in F} \{\phi(N(f))\}.$$

Since H_1 consists of p elements, then $N(F)$ has at most p^n elements. Thus, $k \leq p^n$ and

$$\forall f \in F, \qquad \exists i \in [1,k]\text{: } \text{dist}_G(f,g_i) < b.$$

This means that $F \subset \bigcup_{i=1}^{k} J(g_i,b)$ and therefore

$$H^{-1}(p^n,F) \leq b.$$

Since b can be arbitrarily close to $r(N)$ and N is arbitrary, we get

$$r^a(n) \geq H^{-1}(p^n,F). \tag{4.15}$$

We now prove the reverse inequality. Let $a > H^{-1}(p^n,F)$. Then there exist different elements $g_1,\ldots,g_k \in G$, $k \leq p^n$, such that

$$F \subset \bigcup_{i=1}^{k} J(g_i,a).$$

Let P be a one-to-one mapping of $\{g_1,\ldots, g_k\}$ into H_1^n. Define the information operator

$$N(f) = Pg_i \tag{4.16}$$

and the algorithm

$$\phi(N(f)) = P^{-1}N(f) = g_i, \tag{4.17}$$

where $f \in J(g_i,a)$. Then $r(N) \leq e(\phi,N) \leq a$. Since a can be arbitrarily close to $H^{-1}(p^n,F)$, we have

$$r^a(n) \leq H^{-1}(p^n,F).$$

Due to (4.15), this completes the proof of (4.13).

Due to (4.13), we have $m^a(\epsilon) = \min\{n: H^{-1}(p^n,F) \leq \epsilon\}$. From (4.12) we get

$$m^a(\epsilon) = \min\{n: \exists g_1,\ldots, g_{p^n} \in G, \forall\delta > 0, F \subset \bigcup_{i=1}^{p^n} J(g_i,\epsilon + \delta)\} =$$

$$= \min\{n: \log p^n \geq H(\epsilon,F)\} = \left\lceil \frac{H(\epsilon,F)}{\log p} \right\rceil .$$

which proves (4.14). ■

Note that the information (4.16) is an nth optimal information operator and the algorithm (4.17) is an optimal error algorithm. We discuss the computation of $\phi(N(f))$. If one agrees to code the element $\phi(N(f)) = g_i$ by its unique representation $Pg_i = N(f)$, then the combinatory complexity of ϕ is zero. Thus we have

COROLLARY 4.1 The information (4.16) is nth optimal and the algorithm ϕ defined by (4.17) is an optimal error and optimal complexity algorithm for the class $\Psi^a(n)$ and

$$\text{comp}(\phi) = \text{comp}(\Psi^a,\epsilon) = c_1 m^a(\epsilon) = c_1 \left\lceil \frac{H(\epsilon,F)}{\log p} \right\rceil .$$ ■

5.5 CLASSES OF ALGORITHMS

We comment on the three classes $\Phi(N_\rho)$, $R(N_\rho)$ and $R(N_{\rho,E})$ of algorithms which we have defined in this work. Note that all three consist of algorithms that use an approximate information operator N_ρ.

The class $\Phi(N_\rho)$ of idealized algorithms consists of *all* algorithms that use N_ρ and is independent of the model of computation. The radius $r(N_\rho)$ measures uncertainty in this class. If $r(N_\rho) > \epsilon$, then the approximate information N_ρ does not supply enough information to find an ϵ-approximation, no matter which model of computation might be used.

The class $R(N_\rho)$ of realizable algorithms is a subclass (in general, proper) of $\Phi(N_\rho)$. This class may be independent of a model of computation. It consists of algorithms that have some "structure" properties, e.g., that are linear, polynomial, rational, stable, or on-line algorithms. The radius $r(R(N_\rho),N_\rho)$, which is not less than $r(N_\rho)$, measures uncertainty in $R(N_\rho)$. If $r(R(N_\rho),N_\rho)$

$> \epsilon$, then it is impossible to find an ϵ-approximation by realizable algorithms that use N_ρ.

The class $R(N_\rho,\epsilon)$ of permissible algorithms is a subclass of $R(N_\rho)$. This class does depend on a model of computation and consists of algorithms which yield an ϵ-approximation with finite cost. Note that a realizable algorithm might not be permissible or yield an ϵ-approximation. Therefore, in general, $R(N_\rho,\epsilon)$ is a proper subclass of $R(N_\rho)$ even for $\epsilon = +\infty$.

We can, however, achieve equality among three classes: $\Phi(N_\rho)$, $R(N_\rho)$, and $R(N_\rho,+\infty)$, for $R(N_\rho) = \Phi(N_\rho)$ and a specially defined model of computation. Namely, let any mapping ϕ from $N_\rho(F)$ into G be a primitive operation with unit complexity. Then any idealized algorithm is realizable, and any realizable algorithm is permissible. This yields equality.

Equality between $R(N_\rho)$ and $R(N_\rho,+\infty)$ can also be achieved for any model of computation with a specially constructed R. Namely, let

$$R(N_\rho) = \{\phi\colon \phi(N_\rho(f)) \text{ can be computed at finite cost, } \forall f \in F\}.$$

Then any realizable algorithm is also permissible, which yields equality. Note that $R(N_\rho)$ now depends on a model of computation.

CHAPTER SIX

APPLICATIONS

6.1 INTRODUCTION

A variety of application areas shows how the general theory may be applied to both discrete and continuous problems. Our selection of applications was somewhat arbitrary; other interesting applications could have been chosen. We anticipate that we and others will report of many additional applications in future work.

Since the applications are independent of each other, the reader can pick and choose according on interests.

In the first three applications we study optimal information; there is no concept of complexity. The next three applications have results on both optimal information and complexity.

Sections 6.6 and 6.7 each consist of three subsections. The beginnings of these sections may be consulted for brief summaries of their contents.

The three examples of Section 6.8 (database security, Boolean functions, linear equations) involve finding an information operator whose radius is zero.

We confine ourselves in this work, with a single exception, to the study of the worst case. That exception occurs in Section 6.9 where we discuss typical problems from information theory and decision theory for an average case model. Our motivation for this exception is that we want to indicate how information theory and decision theory fit into our framework. It is our intention to devote substantial future effort to various probabilistic analyses.

6.2 ALGEBRAIC CODING THEORY

Let $f = [f_1, f_2, \ldots, f_k]$ with $f_i \in \{0,1\}$ be a sequence of binary digits (a message) which we want to transmit across a noisy channel. Our problem is to transmit additional check digits such that we can detect the channel errors and correct them so that the exact message can be recovered. In our terminology this problem is formulated as follows.

Let $F = G = \{0,1\}^k$. For $f \in F$ define

$$S(f,\epsilon) = \{f\}, \qquad \forall \epsilon \geq 0. \tag{2.1}$$

Let $H = \{0,1\}^p$ for some nonnegative integer p. Define the exact information operator as

$$N(f) = [g_1(f), g_2(f), \ldots, g_p(f)], \tag{2.2}$$

where $g_i : F \to \{0,1\}$. The function g_i represents how f is coded. As we mentioned in Example 3.2 of Chapter 1, we have

$$r(N) = \begin{cases} 0 & \text{if } S(V(N,f)) \text{ is a singleton, } \forall f \in F, \\ +\infty & \text{otherwise,} \end{cases}$$

where $S(V(N,f)) = \{\tilde{f} \in F : N(\tilde{f}) = N(f)\}$. Thus $S(V(N,f))$ is a singleton iff N is one-to-one. This implies that $\text{card}(N(F)) = \text{card}\, F = 2^k$. Since $N(F) \subset H$ and $\text{card}(H) = 2^p$, we conclude that the necessary condition to make $r(N) = 0$ is

$$p \geq k. \tag{2.3}$$

We define the information error operator E as

$$E(h,\rho) = \{g \in H : \text{HAM}(h,g) \leq \rho\}, \tag{2.4}$$

where $\text{HAM}(h,g)$ equals the number of components in which h and g differ. Let N_ρ be an approximate information operator. We seek exact information of the form (2.2) with minimal p such that $r(N_{\rho,E}) = 0$. That is, we wish to recover the exact message f from the approximate information $N_\rho(f)$.

For simplicity we present the solution of this problem only for $\rho = 1$, i.e., at most one bit can be erroneously transmitted. The general case and many important related problems can be found in a huge literature devoted to this subject (see, for instance, Berlekamp [68]).

Let $H_1 = \{0,1\}$ and let $\Psi(H_1)$ be the class of *all* functions g from F to H_1. In this case, as we proved in Chapter 4, nonadaptive information is as powerful as adaptive, and we therefore need consider only nonadaptive information of the form (2.2).

Suppose that an approximate information operator N_1 has zero radius, $r(N_{1,E}) = 0$. This means that, knowing $N_1(f)$, which belongs to $E(N(f),1)$, we can uniquely identify f. Note that $N_1(f)$ must be one of $p + 1$ vectors. It can be equal to $N(f)$ or to

$$N(f) + [0,\ldots, \underset{i}{1}, 0,\ldots, 0], \qquad i = 1, 2,\ldots, p,$$

where "+" means binary addition, i.e., $0 + 0 = 1 + 1 = 0$ and $1 + 0 = 0 + 1 = 1$.

Thus $p + 1$ elements have to correspond to one element f. Since f is one of the 2^k elements, we have

$$2^k \leq \frac{\mathrm{card}(N_1(F))}{p+1} \leq \frac{2^p}{p+1}.$$

Setting $p = k + s$, where s denotes the total number of check digits, we obtain

$$k + s + 1 \leq 2^s. \tag{2.5}$$

Does there exist an information operator N of cardinality $k + s$ with $k + s + 1$ as close to 2^s as possible and $r(N_{1,E}) = 0$? The answer is yes and N is defined as follows. Let

$$g_i(f) = f_i, \qquad i = 1, 2, \ldots, k. \tag{2.6}$$

To define $g_{k+1}(f), \ldots, g_{k+s}(f)$, let

$$K = [K_1, I] \tag{2.7}$$

be an $s \times p$ matrix, where I is the $s \times s$ identity matrix and K_1 is an $s \times k$ matrix chosen so that all elements of K_1 are zero or one and all columns of K are distinct and nonzero. It is obvious that such a matrix exists iff $p \leq 2^s - 1$, i.e., iff (2.5) holds. Define

$$\begin{bmatrix} g_{k+1}(f) \\ \vdots \\ g_{k+s}(f) \end{bmatrix} = K_1 \begin{bmatrix} f_1 \\ \vdots \\ f_k \end{bmatrix}. \tag{2.8}$$

Note that g_{k+i} depends linearly on f and

$$K[f_1, \ldots, f_k, g_{k+1}(f), \ldots, g_{k+s}(f)]^T = KN(f)^T = 0.$$

The linear code defined by the matrix K is called a *Hamming code.*

We show that for the information operator N defined by (2.6) and (2.8) we have $r(N_{1,E}) = 0$. Indeed, let $N_1(f)$ be approximate information. Then

$$N_1(f) = N(f) + [0, \ldots, \underset{i}{1}, \ldots, 0] = N(f) + e_i \qquad \text{or} \qquad N_1(f) = N(f).$$

Let $y = KN_1(f)^T$. Hence $y = 0$ or $y = Ke_i^T$, i.e., y is the ith column of K. Since all columns of K are distinct and nonzero, then knowing y we can find which bit was erroneously transmitted. More precisely, if $y = 0$, then $N_1(f) = N(f)$

and the first k components of $N(f)$ yield f. Otherwise y is the ith column of K which indicates that the ith bit was erroneously transmitted and $N(f) = N_1(f) - e_i$ is the correct information and f is obtained as the first k components of $N(f)$. In either case, we know f, which implies $r(N_{1,E}) = 0$.

What is the minimal cardinality $n = k + s$ for which $r(N_{1,E}) = 0$? From (2.5) it follows that the minimal cardinality number $s(k)$ is equal to the smallest s for which $k + s + 1 \leq 2^s$. It is easy to check that $s(1) = 2$, $s(2) = 3$, and

$$s(k) = \lceil \log_2 k \rceil + a, \qquad \forall k \geq 3,$$

where $a = 0$ or $a = 1$. Thus we have

THEOREM 2.1 For the algebraic coding problem (2.1), the cardinality number $m^a(1,0)$ is given by

$$\mathrm{m}^a(1,0) = m^{\mathrm{non}}(1,0) = \min\{s: k + s + 1 \leq 2^s\} \simeq \lceil \log k \rceil .$$

The information operator N defined by (2.2), (2.6), and (2.8), which is a Hamming code, is an $m(1,0)$th optimal approximate operator. ■

6.3 DISTRIBUTED COMPUTATION

We now show how a distributed computation problem fits into our model. Our presentation is primarily based on Abelson [80]. Related problems are considered by Abelson [78], Gentleman [78], Thompson [80] and Yao [79].

We consider the computation of the values of a smooth function that depends on large amount of data. We assume that the data are partitioned into disjoint subsets and a processor is assigned to each subset of data. Each processor can compute some function of its input and transfer its results to other processors. The problem is to find or to bound from below the total information transfer, i.e., the number of values which must be transmitted between processors to compute the value of a smooth function.

We formulate this problem in our terminology. Let $F = X \times Y$, where $X \subset \mathbb{R}^n$ and $Y \subset \mathbb{R}^m$ are open subsets. Without loss of generality, we assume that $n \leq m$. By C^2 we mean the class of functions from F into $\mathbb{R}$ whose second derivative is continuous. Let

$$S : F \to \mathbb{R} \tag{3.1}$$

belong to C^2. Define

$$S(f,\epsilon) = \{S(f)\}, \qquad f = (x,y) \in F. \tag{3.2}$$

Let $H_1 = \mathbb{R}$. We define the class $\Psi(\mathbb{R})$ of primitive information operators as the subset of C^2 which consists of functions $U : F \to \mathbb{R}$ of the form

$$U(f) = U(x,y) = H(x),$$

or

$$U(f) = U(x,y) = g(y)$$

for some functionals h and g. That is, a primitive information operator is a smooth functional, in general nonlinear, that depends only on x or y, i.e., only on a part of the data.

We ask what is the minimal cardinality of an adaptive information operator for which we can solve our problem. Since the solution operator S defined by (3.2) does not depend on ϵ, this question is equivalent to finding the cardinality number. (We assume $R = \Phi(N)$ and $\rho = 0$ here.)

Consider, for example, the information operator $N = I$, i.e.,

$$N(f) = f = [x_1,\dots, x_n, y_1,\dots, y_m].$$

Then the algorithm $\phi(N(f)) = \phi(f) = S(f)$ has radius equal to zero and belongs to the idealized class $\Phi(N)$. Thus the cardinality number satisfies

$$m^a(0) \le n + m. \tag{3.3}$$

Note that $\text{card}(I) = n + m$ and $t = \text{card}(I) - m$ can be interpreted as the total information transfer between two processors. That is, the processor that is assigned to x transmits all the x_i to the second processor assigned to y. The second processor, which now knows x and y, computes $S(f)$ and solves the problem. The number of values transmitted between two processors is n. Can we do better? That is, do we have to transmit n values?

Abelson [80] proved the following theorem. Let $\text{rank}(f)$ be the rank of the $n \times m$ Hessian matrix $M = (\partial^2 S(f) / \partial x_i \partial y_j)$. Let N^a be an adaptive information operator such that $r(N^a, f_1) = 0$, where f_1 belongs to a neighborhood of f. Let $t(N^a)$ be the total information transfer needed to compute $N^a(f_1)$. Then

$$\text{card}(N^a) \ge t(N^a) \ge \text{rank}(f). \tag{3.4}$$

Thus, (3.3) and (3.4) imply that

$$\text{rank}(f) \le m^a(0) \le m + n, \qquad \forall f \in F. \tag{3.5}$$

The essence of Abelson's theorem is that for many practical cases

$$\text{rank}(f) = n \qquad \text{and} \qquad \text{card}(N^a) = t(N^a) + m,$$

which yields $m^a(0) = m + n$.

If rank$(f) = n$, then all the data has to be transmitted in order to compute the values of a smooth function, and this can be done by simply transmitting all x_i. For instance, let $n = m$ and

$$g(f) = \sum_{i=1}^{n} x_i y_i.$$

Then rank $(f) = n$ and the transfer of all data is necessary.

Finally we want to comment on the smoothness assumption for primitive information operators. Without some assumption, the problem becomes trivial since one could transmit all n numbers x_i by one nonsmooth nonlinear functional and the total information transfer becomes one.

6.4 CONTINUOUS BINARY SEARCH

In this section we exhibit the optimal approximate information for the continuous binary search problem. Recall that many concepts in the previous chapters were illustrated by the binary search problem for the discrete case. The presentation is based primarily on Rivest, Meyer, Kleitman, Winklmann, and Spencer [80].

Let $F = (0,1]$. Let $G = 2^F$ be the class of all subsets of F. Let μ be the Lebesgue measure on $\mathbb{R}$. Define

$$S(f,\epsilon) = \{A \in G: f \in A \text{ and } \mu(A) \leq \epsilon\}. \tag{4.1}$$

Thus, an ϵ-approximation is a set that contains f and whose measure is at most ϵ. Let $H_1 = \{0,1\}$ and $\Psi(H_1)$ be the class of primitive information operators U such that

$$U(f) = \begin{cases} 1 & \text{if } f \in T, \\ 0 & \text{if } f \notin T, \end{cases} \tag{4.2}$$

where T is a measurable subset, $T = T(U)$. Define the information error operator E as

$$E(h,k) = \{g \in H_1^n : \mathrm{HAM}(h,g) \leq k\}$$

for a given nonnegative integer k.

We seek an nth optimal approximate information, the nth minimal radius, and the ϵ-cardinality number (for the classes of idealized algorithms). We define an information operator N, $n \geq k$, of the form

$$N(f) = [Q(f,c_1), Q(f,c_2),\ldots, Q(f,c_n)], \tag{4.3}$$

where $Q(f,c) = 1$ if $f \leq c$ and 0 otherwise. The numbers c_i belong to F and the choice of c_i depends on the previous answers $Q(f,c_1),\ldots, Q(f,c_{i-1})$. Thus N is adaptive. To define c_i, we proceed as follows. Let $c_1 = 1/2$. Assume inductively that $c_1,\ldots, c_{i-1}$ are defined. Let $q = n - i + 1$ and define

$$A_q = (A_q^0, A_q^1,\ldots, A_q^k),$$

where A_q^j is a subset of F such that $f \in A_q^j$ iff exactly j of the previous answers were incorrect. (Note that $A_n = (F, \emptyset,\ldots, \emptyset)$.) As in Rivest et al. [80], denote

$$\left(\binom{n}{m}\right) = \binom{n}{0} + \binom{n}{1} + \cdots + \binom{n}{m},$$

$$w(q,A_q) = \sum_{j=0}^{k} \left(\binom{q}{k-j}\right) \mu(A_q^j).$$

Let T be a measurable subset of $\mathbb{R}$. Knowing A_q, define

$$^yA_{q-1} = (^yA_{q-1}^0,\ldots, {}^yA_{q-1}^k),$$

$$^NA_{q-1} = (^NA_{q-1}^0,\ldots, {}^NA_{q-1}^k),$$

where

$$^yA_{q-1}^0 = A_q^0 \cap T, \qquad {}^NA_{q-1}^0 = A_q^0 - T,$$

and for $j = 1, 2,\ldots, k$,

$$^yA_{q-1}^j = (A_q^j \cap T) \cup (A_q^{j-1} - T),$$

$$^NA_{q-1}^j = (A_q^j - T) \cup (A_q^{j-1} \cap T).$$

We define c_i such that for $T = (0,c_i]$ we have

$$w(q - 1, {}^yA_{q-1}) = w(q - 1, {}^NA_{q-1}).$$

Such a c_i exists due to properties of w and $^yA_{q-1}$ and $^NA_{q-1}$.

THEOREM 4.1 (Rivest et al. [80]) The information operator (4.3) is an nth approximate optimal information (for $\Psi^a(n)$ and k) and

$$r(N_{k,\epsilon}) = r^a(n,k) = \left(\binom{n}{k}\right) 2^{-n},$$

$$m^a(k,\epsilon) = \min\{n: \left(\binom{n}{k}\right) 2^{-n} \leq \epsilon\} \simeq \log \tfrac{1}{\epsilon} + k \log\log \tfrac{1}{\epsilon}. \qquad \blacksquare$$

This shows that if k is small relative to n, we don't lose much from the case of no errors. Note that the nth approximate information $N(f)$ yields the final set $A = \cup_{j=0}^{k} A_0^j$ containing f. It is easy to observe that A is a union of disjoint intervals. The further study of this problem for continuous and discrete cases, as well as a relation to the problem of finding the minimum root of a set of increasing functions, may be found in Rivest et al. [80]. The discrete binary search problem will be continued in Appendix A.

6.5 BIN PACKING

Let $f = [f_1, f_2, \ldots, f_n]$ be a given sequence of real numbers, $f_i \in [a,1]$, where a is a small positive number. Let $\text{BIN}_1, \text{BIN}_2, \ldots$ be a sequence of at most n bins, each of unit capacity. In the bin-packing problem we assign each f_i to a bin in such a way that the sum of numbers in each bin does not exceed one and the total number of bins used is minimal (see, for instance, Garey and Johnson [79], Johnson [74], and Yao [80]).

We formalize this problem in our terminology. Let $F = [a,1]^n$ and let

$$k(f) = [k_1(f), k_2(f), \ldots, k_n(f)] \tag{5.1}$$

be a possible packing of f, that is, $k_i(f)$ is the number of the bin to which f_i is assigned and

$$\sum_{i:k_i(f)=b} f_i \leq 1, \tag{5.2}$$

with $b = k_j(f)$, $j = 1, 2, \ldots, n$. Equation (5.2) guarantees that the sum of numbers in the bth bin does not exceed one. Let G be the set of all vectors $k(f)$ satisfying (5.1) and (5.2). Let

$$\| k(f) \| = \max_{1 \leq i \leq n} k_i(f). \tag{5.3}$$

If there are no gaps in the packing, then $\| k(f) \|$ is the total number of bins used by the packing $k(f)$. For a given f, define $k^*(f) = [k_1^*(f), \ldots, k_n^*(f)] \in G$ such that

$$\| k^*(f) \| = \min_{k(f) \in G} \| k(f) \| . \tag{5.4}$$

Thus $k^*(f)$ is the packing that minimizes the total number of bins used. From (5.2) it follows that

$$\| k^*(f) \| \geq \left\lceil \sum_{i=1}^{n} f_i \right\rceil \geq na. \tag{5.5}$$

Thus $\| k^*(f) \|$ goes to infinity with n. Define the solution operator as

$$S(f,\epsilon) = \{g(f) \in G: \frac{\| g(f) \|}{\| k^*(f) \|} \leq 1 + \epsilon\}, \tag{5.6}$$

that is, as the set of all possible packings that use at most $(1 + \epsilon)$ times more bins than the optimal one. Let $N(f) = f$ and $\rho = 0$. Since N is complete, $r(N) = 0$.

We now define the set P of primitive operations as comparison and the four arithmetic operations, each with unit complexity. What is the ϵ-complexity of the bin-packing problem as a function of n? It is known that the bin-packing problem is NP-complete and therefore it is very likely that the complexity ($\epsilon = 0$) of the bin-packing problem is an exponential function of n. However, if ϵ is large enough, there exist algorithms that find an ϵ-approximation whose complexity is a polynomial of small degree in n. We report a number of known algorithms.

We consider two classes of realizable algorithms. The first one is the class $\Phi(N)$ of idealized algorithms, i.e., $R = \Phi(N)$. Sometimes $\Phi(N)$ is called the class of *off-line algorithms*. The second one is the class R_1 of *on-line algorithms* defined as follows. Let

$$R_1 = \{\phi: \phi(f) = [\phi_1(f),\ldots, \phi_n(f)] \in G \text{ and } \phi_i(f) \text{ depends only on } f_1, f_2,\ldots, f_i\}. \tag{5.7}$$

That is, an algorithm ϕ is on line if the numbers $f_1, \ldots, f_n$ are available one at a time and ϕ has to assign a bin to f_i before the next numbers $f_{i+1},\ldots, f_n$ become available. Recently, Yao [80] showed that

$$r(R_1,N) \geq \tfrac{1}{2}\,[1 + o(1)], \quad \text{as } n \to \infty. \tag{5.8}$$

Thus, it is impossible to find an ϵ-approximation that uses an on-line algorithm with $\epsilon < 1/2$ and large n.

For large ϵ, there exist algorithms that produce an ϵ-approximation with small complexity. For instance, the first fit (FF) algorithm assigns f_j sequentially to the bin with the smallest possible i, that is, the smallest i whose current content does not exceed $1 - f_j$. Note that the FF algorithm is on-line. It is known that

$$e(\text{FF},N) = \tfrac{7}{10}\,[1 + o(1)], \quad \text{as } n \to \infty. \tag{5.9}$$

and the complexity is

$$\text{comp(FF)} = \Theta(n \log n).$$

Recently, Yao [80] improved (5.9) by proposing an on-line algorithm RFF with

$$e(\mathrm{RFF}, N) = \tfrac{2}{3}\,[1 + o(1)], \quad \text{as } n \to \infty, \qquad (5.10)$$

$$\mathrm{comp}(\mathrm{RFF}) = \Theta(n \log n).$$

Thus if $\epsilon > 2/3$ and n is large, then (5.10) yields

$$\mathrm{comp}(R_1, N, \epsilon) = 0(n \log n). \qquad (5.11)$$

This gives an upper bound; $\mathrm{comp}(R_1,N,\epsilon)$ is unknown. The ϵ-complexity of the bin-packing problem is also unknown if $\epsilon \in [1/2, 2/3]$.

We now consider an algorithm that belongs to the class $\Phi(N)$ of idealized (off-line) algorithms. This is the first-fit-decreasing (FFD) algorithm which first sorts $(f_1, \ldots, f_n)$ into decreasing order and then performs the first-fit algorithm. It is known that

$$e(\mathrm{FFD}, N) = \tfrac{2}{9}\,[1 + o(1)], \quad \text{as } n \to \infty, \qquad (5.12)$$

$$\mathrm{comp}(\mathrm{FFD}) = \Theta(n \log n). \qquad (5.13)$$

Thus if $\epsilon > 2/9$ and n is large, then (5.12) and (5.13) yield

$$\mathrm{comp}(N, \epsilon) = 0(n \log n). \qquad (5.14)$$

This gives an upper bound; $\mathrm{comp}(N,\epsilon)$ is unknown. The ϵ-complexity of the bin packing problem is also unknown if $\epsilon \le 2/9$. Yao [80] improved (5.11) by presenting an algorithm RFFD with error no larger than $(\frac{2}{9} - \delta)$ for some positive δ.

We stress once more that for $\epsilon = 0$, the bin-packing problem is NP-complete and its complexity is likely to be an exponential function of n. Thus the ϵ-complexity behaves very irregularly, jumping from a low-degree polynomial for moderate ϵ to (probably) an exponential function for $\epsilon = 0$. There are other problems with the same characteristics, such as the traveling-salesman problem. The reader may find related material in Garey and Johnson [79].

6.6 NONLINEAR EQUATIONS

In this section we deal with the approximate solution of the scalar or multivariate nonlinear equation $f(x) = 0$. We consider a number of classes of f and two different approximation criteria. The first criterion is to find an element x which is close to an exact solution α, $f(\alpha) = 0$. The second criterion is to find an element x for which $\| f(x) \|$ is small. We consider adaptive and nonadaptive information operators. We shall show that the strength of adaption relative to nonadaption depends strongly on the class of functions.

(i) Bisection Algorithm

In this subsection we present the optimal properties of the bisection algorithm for the approximate solution of a scalar nonlinear equation. Let

$$F = \{f: f: [a,b] \to \mathbb{R}, f \text{ is continuous}, f(a) \le 0,\ \ f(b) \ge 0 \text{ and } f \text{ has exactly one zero}\}. \tag{6.1}$$

Let $G = [a,b]$ and

$$S(f,\epsilon) = \begin{cases} \{x \in G: |x - f^{-1}(0)| < \epsilon\} & \text{if } \ \epsilon > 0, \\ \{f^{-1}(0)\} & \text{if } \ \epsilon = 0. \end{cases} \tag{6.2}$$

Let $H_1 = \mathbb{R}$ and $\Psi(H_1)$ be the class of primitive information operators U, where U is a real linear functional. Consider first the class $\Psi^{\text{non}}(n)$ of nonadaptive information operators with cardinality at most n. Let

$$N^{\text{non}}(f) = [f(x_1), f(x_2), \ldots, f(x_n)], \tag{6.3}$$

where $x_i = a + i(b - a)/(n + 1)$, $i = 1, 2, \ldots, n$. Then $N^{\text{non}} \in \Psi^{\text{non}}(n)$. Assuming $R = \Phi(N)$ and $\rho = 0$, Traub and Woźniakowski [80a, pp. 165–168] proved that N^{non} is an nth optimal information operator in the class $\Psi^{\text{non}}(n)$ and

$$r(N^{\text{non}}) = r^{\text{non}}(n) = \frac{b - a}{2(n + 1)} . \tag{6.4}$$

From (6.4) we obtain the ϵ-cardinality number as

$$m^{\text{non}}(\epsilon) = \left\lfloor \frac{b - a}{2\epsilon} \right\rfloor . \tag{6.5}$$

To derive the ϵ-complexity, assume that the set P of primitive operations consists of function evaluations with complexity equal to c, comparison, and the four arithmetic operations, each with unit complexity. Let $\Psi^{\text{non}} = \bigcup_{n=1}^{\infty} \Psi^{\text{non}}(n)$.

We now present an optimal complexity algorithm for N^{non} defined by (6.3). Let $j = j(N^{\text{non}})$ be the index such that

$$f(x_j) \le 0 \qquad \text{and} \qquad f(x_{j+1}) \ge 0$$

($x_0 = a$ and $x_{n+1} = b$). Define

$$\phi(N^{\text{non}}(f)) = \begin{cases} (x_j + x_{j+1})/2 & \text{if } \ f(x_j)\,f(x_{j+1}) \ne 0, \\ x_i & \text{if } \ f(x_i) = 0 \quad (i = j \text{ or } i = j+1). \end{cases}$$

Then ϕ is a strongly optimal error algorithm with combinatory complexity proportional to log n. The algorithm ϕ is also a nearly optimal complexity algorithm for the class Ψ^{non} and

$$\operatorname{comp}(\Psi^{\text{non}},\epsilon) = \operatorname{comp}(\phi)\,[1+o(1)] = [c+o(1)] \left\lfloor \frac{b-a}{2\epsilon} \right\rfloor . \tag{6.6}$$

Thus the ϵ-complexity in the class of nonadaptive information operators is proportional to $1/\epsilon$ as ϵ goes to zero.

We now consider the class $\Psi^a = \bigcup_{n=1}^{\infty} \Psi^a(n)$ of adaptive information operators with finite cardinality. Let

$$N^a(f) = [f(y_1), f(y_2), \ldots, f(y_n)], \tag{6.7}$$

where y_i are inductively defined as follows. Let $a_0 = a$ and $b_0 = b$. Let $[a_j, b_j]$ denote the interval containing the zeros of the functions from F, which agree with f at $y_1, y_2, \ldots, y_j$, where $j = 0, 1, \ldots, i-1$. Then

$$y_i = \tfrac{1}{2}(a_{i-1} + b_{i-1}),$$

$$a_i = \begin{cases} a_{i-1} & \text{if } f(y_i) > 0, \\ y_i & \text{if } f(y_i) \le 0, \end{cases} \qquad b_i = \begin{cases} b_{i-1} & \text{if } f(y_i) < 0, \\ y_i & \text{if } f(y_i) \ge 0. \end{cases}$$

The bisection algorithm ϕ^b uses the adaptive information N^a and is defined as

$$\phi^b(N^a(f)) = \frac{a_n + b_n}{2} . \tag{6.8}$$

It is known that the bisection algorithm is a strongly optimal error algorithm and

$$e(\phi^b) = r(N^a) = (b-a)2^{-(n+1)}.$$

Note that $r(N^a)$ is significantly smaller than the nth minimal radius of information for the class of nonadaptive information operators (see (6.4)). Thus adaption is much more powerful than nonadaption for this problem.

In GTOA (p. 170) we conjecture that the information N^a defined by (6.7) is nth optimal information in the class $\Psi^a(n)$, i.e.,

$$r^a(n) = \inf_{N \in \Psi^{\text{non}}(n)} r(N) = r(N^a) = (b-a)2^{-(n+1)}.$$

The correctness of this conjecture was established recently by Sikorski [81]. (The information (6.7) is also optimal in the class $\Psi^a(n)$ if f is infinitely differentiable and the zero of f is simple.) We easily conclude that

$$m^a(\epsilon) = \left\lfloor \log\left(\frac{b-a}{\epsilon}\right) \right\rfloor .$$

Compare with (6.5) for the class of nonadaptive information.

We derive the ϵ-complexity assuming the same set P of primitive operations as for the nonadaptive case. Once a_n, b_n are known, the combinatory complexity of the bisection algorithm ϕ^b is only two. Since the information complexity of N^a for $n = m^a(\epsilon)$ is $(c_1+3)m^a(\epsilon) - 1$, we obtain

$$\text{comp}(\phi^b) = (c_1 + 3) \left\lfloor \log\left(\frac{b-a}{\epsilon}\right) \right\rfloor + 1 .$$

Since N^a is nth optimal, we easily conclude that

$$\text{comp}(\Psi^a,\epsilon) = (c_1 + a_1) \left\lfloor \log\left(\frac{b-a}{\epsilon}\right) \right\rfloor + a_2, \tag{6.9}$$

where $a_1 \in [0,3]$ and $a_2 \in [0,1]$. Since c_1 is usually large, the bisection algorithm is a nearly optimal complexity algorithm for the class Ψ^a.

Note that (6.6) and (6.9) can be rewritten as

$$\text{comp}(\Psi^{\text{non}},\epsilon) = \Theta\left(\tfrac{1}{\epsilon}\right), \qquad \text{comp}(\Psi^a,\epsilon) = \Theta\left(\log \tfrac{1}{\epsilon}\right).$$

This exhibits the power of adaptive information for this problem.

(ii) Lipschitz Class

In this subsection we present the results of a recent paper of Sikorski [80] who studies the following problem. Let $G = [0,1]^d$ and let $f\colon G \to \mathbb{R}^k$. Assume that

$$f(x) = 0 \tag{6.10}$$

has a solution and that f satisfies a Lipschitz condition with a given positive constant K, i.e.,

$$\| f(x) - f(y) \| \le K \| x - y \|, \tag{6.11}$$

where $\| \cdot \|$ denotes the infinity norm in the appropriate space, i.e.,

$$\| x \| = \max_{1 \le i \le d} | x_i | .$$

We seek a vector x with small residual vector. More precisely, we define an ϵ-approximation as a vector x such that

$$\| f(x) \| \leq \epsilon. \tag{6.12}$$

Note that without loss of generality we can assume that $k = 1$. Indeed, let

$$g(x) = \max_{1 \leq j \leq k} | f_j(x) |.$$

Then g satisfies (6.11) with the same constant K as f, g has the same zeros as f, and ϵ-approximations for g and f are also the same. Therefore we assume that $k = 1$ and define

$$F = \{f: f: G \to \mathbb{R}, f \text{ satisfies (6.11) and } f(x) = 0 \text{ has a solution}\}. \tag{6.13}$$

The solution operator is given by

$$S(f,\epsilon) = \{x \in G: | f(x) | \leq \epsilon\}. \tag{6.14}$$

Note that for $\epsilon = 0$, $S(f,0)$ consists of all solutions of the equation $f(x) = 0$ and need not be a singleton.

Let $H_1 = \mathbb{R}$ and $\Psi(H_1)$ be the class of all primitive information operators U such that

$$U(f) = f(x) \tag{6.15}$$

for some $x = x(U) \in G$. That is, one function evaluation is a primitive information operator. We wish to find nth optimal information operators (assuming that $R = \Phi(N)$ and $\rho = 0$).

Suppose for a moment that n is of the form

$$n = M^d - 1 \tag{6.16}$$

for some integer M, $M > 1$. Define the set X as

$$X = \left\{x \in G: x = \left[\frac{2j_1 - 1}{2M}, \ldots, \frac{2j_d - 1}{2M}\right]\right\} \tag{6.17}$$

for $j_i = 1, 2, \ldots, M$ and $i = 1, 2, \ldots, d$. The set X has $M^d = n + 1$ elements. Let $x_1^*, \ldots, x_{n+1}^*$ be distinct elements of X, that is, $X = \{x_1^*, \ldots, x_{n+1}^*\}$. For fixed $i \in [1, n + 1]$ define the nonadaptive information

$$N^{\text{non}}(f) = [f(x_1^*), \ldots, f(x_{i-1}^*), f(x_{i+1}^*), \ldots, f(x_{n+1}^*)]. \tag{6.18}$$

Note that $N^{\text{non}} \in \Psi^{\text{non}}(n)$. Sikorski [80] proves that for each i, N^{non} is an nth optimal information operator in the class $\Psi^{a}(n)$ of adaptive information operators and

$$r(N^{\text{non}}) = r^{\text{non}}(n) = r^{a}(n) = \frac{K}{2M} \,. \tag{6.19}$$

(In fact, he also proves that (6.18) is almost optimal if arbitrary linear functionals are permitted as primitive information operators.) Thus, unlike in subsection (i), adaption is of no help for this problem. Choose k such that

$$|\, f(x_k^*) \,| = \min\{|f(x_j^*)| : j \in [1,n+1] \text{ and } j \neq i\}.$$

Define

$$\phi(N^{\text{non}}(f)) = \begin{cases} x_i^* & \text{if } |\, f(x_k^*) \,| > K/(2M), \\ x_k^* & \text{otherwise.} \end{cases} \tag{6.20}$$

Then ϕ is an optimal error algorithm, but not a strongly optimal error algorithm.

Let the set P of primitive operations consist of function evaluations with complexity c, and comparison and the four arithmetic operations with unit complexity. Since the computation of ϕ requires only n comparisons, its combinatory complexity is n. Sikorski [80] also derives a strongly optimal error algorithm but conjectures that its combinatory complexity is exponential.

For the general case, i.e., if n does not satisfy (6.16), Sikorski [80] proves that

$$\begin{aligned} m^{a}(\epsilon) &= m^{\text{non}}(\epsilon)\,[1+o(1)] = \left(\frac{K}{2\epsilon}\right)^{d} [1+o(1)], \\ \text{comp}(\Psi^{a},\epsilon) &= \text{comp}(\Psi^{\text{non}},\epsilon)\,[1+o(1)] = (c+1)\left(\frac{K}{2\epsilon}\right)^{d} [1+o(1)], \end{aligned} \tag{6.21}$$

where $\Psi^{a} = \cup_{n=1}^{\infty} \Psi^{a}(n)$ and $\Psi^{\text{non}} = \cup_{n=1}^{\infty} \Psi^{\text{non}}(n)$. Since

$$\text{comp}(\phi) = \text{comp}(\Psi^{\text{non}},\epsilon)\,[1+o(1)],$$

we conclude that ϕ is a nearly optimal complexity algorithm for both Ψ^{a} and Ψ^{non}.

Equation (6.21) states that the ϵ-complexity in the adaptive class Ψ^{a} and in the nonadaptive class Ψ^{non} is essentially the same and is proportional to ϵ^{-d}. This shows how the ϵ-complexity depends on dimensionality.

For $d = 1$, we compare the results of subsections (i) and (ii). For nonadaptive information, the ϵ-complexities for both classes (6.1) and (6.13) are essentially the same. Thus, the change of sign in the class (6.1) supplies information as strong as the Lipschitz condition in the class (6.13). For adaptive information, the ϵ-complexity of the class (6.1) is significantly less than the ϵ-complexity of the class (6.13). Thus, the change of sign now provides much more powerful information than the Lipschitz condition.

More detailed discussion and further results may be found in Sikorski [80]. For $d = 1$, this problem was analyzed by Sukharev [76].

(iii) Polynomial Equations

In this section we report a recent result of Wasilkowski [79] who considers the following problem. Let $G = \mathbb{C}$ and let $x_0 \in G$ be given. Let F_1 be the class of all polynomials with complex coefficients and all simple zeros. Let $S(f)$ be the set of all zeros of f. Define

$$\operatorname{dist}(x_0,S(f)) = \min\{|x_0 - \alpha| : \alpha \in S(f)\}.$$

For a given positive real L define

$$F = F(x_0,L) = \{f \in F_1 : \operatorname{dist}(x_0,S(f)) < L\}. \tag{6.22}$$

The solution operator is given by

$$S(f,\epsilon) = \{x \in G\colon \operatorname{dist}(x,S(f)) \leq \epsilon \operatorname{dist}(x_0,S(f))\}. \tag{6.23}$$

We assume that $0 \leq \epsilon < 1$ since otherwise x_0 belongs to $S(f,\epsilon)$. Consider the adaptive information operator N_n^a of the form

$$N_n^a(f) = [L_1(f), L_2(f;y_1),\ldots, L_n(f;y_1,\ldots, y_n)], \tag{6.24}$$

where $y_i = L_i(f;y_1,\ldots, y_{i-1})$ and L_i depends linearly on its first argument. Wasilkowski [79] proves that ($R = \Phi(N)$ and $\rho = 0$)

$$r(N_n^a) = +\infty, \qquad \forall n. \tag{6.25}$$

This means that it is impossible to find an ϵ-approximation no matter how many evaluations of linear functionals are allowed. Note that (6.25) holds no matter how small L is in (6.22) and how close ϵ is to one. Since most commonly used algorithms depend on an information operator of the form (6.24), this is a very surprising negative result. In any reasonable model of computation, (6.25) implies that the ϵ-complexity is infinity.

One can interpret (6.25) as stating that the class F of polynomials with complex coefficients and all simple zeros is too large. Thus, to make the radius finite, one has to make the class F smaller. Recently, Trojan [80] proved that if the assumption of complex coefficients is replaced by the assumption of rational coefficients, then the opposite result holds. More precisely, let F_0 be the subclass of polynomials from F with rational coefficients and let

$$N(f) = f(\pi). \tag{6.26}$$

Thus N consists of only one function evaluation. Note that

$$V(N,f) = \{\tilde{f} \in F_0: \tilde{f}(\pi) = f(\pi)\} = \{f\},$$

which means that $V(N,f)$ is a singleton and $r(N) = 0$.

Indeed, if $\tilde{f}(\pi) = f(\pi)$, then $(\tilde{f} - f)(\pi) = 0$. This means that π is a zero of a rational polynomial. Since π is not an algebraic number, we have $\tilde{f} - f = 0$. Of course, in this case the class of idealized algorithms is too large and one has to define permissible and realizable algorithms in order to guarantee that an ϵ-approximation can actually be computed.

Note that π in (6.26) can be replaced by any number that is not algebraic, and still the radius of information will be zero.

6.7 LINEAR EQUATIONS

In this section we discuss the approximate solution of linear equations $Ax = b$. In subsection (i) we consider partial information operators, which are widely used for large linear systems, and we assume the real number model of computation. We establish very strong optimal error properties of the minimal residual algorithm (which is a variant of the conjugate gradient algorithm). As we note in Remark 7.1, the minimal residual algorithm performs at most one more step than the optimal algorithm, even when competing with algorithms that know all eigenvalues of the matrix A. In subsection (ii) we consider an approximate complete information operator and the fixed-precision floating-point model. We show how some important roundoff concepts are related to the radius of approximate information. In the final subsection we consider complete information and the variable-precision floating-point model of computation. We study, for various models of computation, the dependence of ϵ-complexity on the dimension of the linear system.

(i) Large Linear Systems

In this subsection we deal with the approximate solution of large linear systems. We consider the information operators commonly used by iterative algorithms. This subsection is based on Traub and Woźniakowski [80b].

Let F be a subclass of the class of $n \times n$ nonsingular real matrices. Let $G = \mathbb{R}^n$ with the l_2 norm, $\| x \| = \sqrt{(x,x)}$. Let b be a given $n \times 1$ real vector such that $\| b \| = 1$. Define the solution operator as

$$S(A,\epsilon) = \{g \in G: \| Ag - b \| \le \epsilon\}, \qquad A \in F. \tag{7.1}$$

Thus, $S(A,\epsilon)$ is the set of vectors g with the residual $\| Ag - b \|$ not exceeding ϵ. Define the information operator

$$N_k(A) = [b, Ab, \ldots, A^k b]. \tag{7.2}$$

We seek the radius of N_k, algorithms that use N_k and have errors comparable to the radius of N_k, and the minimal k such that $\mathrm{r}(N_k) \le \epsilon$. We assume that $R = \Phi(N_k)$ and $\rho = 0$.

We recall the definition of the minimal residual algorithm ϕ_k^{mr}. Define the coefficients $c_1^*, c_2^*, \ldots, c_k^*$ such that

$$\| b - c_1^* Ab - \cdots - c_k^* A^k b \| = \min_{c_i} \| b - c_1 Ab - \cdots - c_k A^k b \|. \tag{7.3}$$

The minimal residual algorithm ϕ_k^{mr} is defined as

$$\phi_k^{mr}(N_k(A)) = c_1^* b + \cdots + c_k^* A^{k-1} b. \tag{7.4}$$

If $\tilde{A} \in V(N_k,A)$, then $\tilde{A}^i b = A^i b$ for $i = 1, 2, \ldots, k$, and therefore $\tilde{A}\phi_k^{mr}(N_k(A)) = A\phi_k^{mr}(N_k(A))$. This yields that the local error of ϕ_k^{mr} is given by

$$e(\phi_k^{mr},A) = \| A\phi_k^{mr}(N_k(A)) - b \|. \tag{7.5}$$

Note that knowing $N_k(A)$ we can compute $e(\phi_k^{mr},A)$. It is obvious that for $k = n$ we have $e(\phi_n^{mr},A) = 0$.

The class F is called orthogonally invariant iff

$$A \in F \Rightarrow QAQ^T \in F \tag{7.6}$$

for every orthogonal matrix Q. From Theorem 4.1 of Traub and Woźniakowski [80b] it easily follows that if F is orthogonally invariant, then

$$e(\phi_{k+1}^{mr},A) \le r(N_k,A) \le e(\phi_k^{mr},A), \qquad \forall A \in F. \tag{7.7}$$

This means that the local radius $r(N_k,A)$ is bounded from below by the local error of the minimal residual algorithm ϕ_{k+1}^{mr} that uses the information N_{k+1}. The bounds (7.7) are sharp, i.e., it can happen that $r(N_k,A) = e(\phi_k^{mr},A)$ for some matrix A and orthogonally invariant class F, and $r(N_k,A) = e(\phi_{k+1}^{mr},A)$ for another choice of A and F.

Let $k(A)$ be the minimal integer such that $r(N_k,A) \le \epsilon$ and $k(F) = \max\{k(A) : A \in F\}$. We call $k(A)$ the optimal matrix index and $k(F)$ the optimal class index. Similarly, let $k(\phi^{mr},A)$ be the minimal integer such that $e(\phi_k^{mr},A) \le \epsilon$ and $k(\phi^{mr}) = \max\,\{k(\phi^{mr},A) : A \in F\}$.

Then (7.7) easily yields

$$k(A) = k(\phi^{mr}, A) + a_1, \qquad k(F) = k(\phi^{mr}) + a_2, \tag{7.8}$$

where $a_i = 0$ or $a_i = -1$, $i = 1, 2$. Thus, with the minimal residual algorithm, we have to perform at most one step more than with a strongly optimal error algorithm. Thus the minimal residual algorithm enjoys a very strong optimal error property although it is not a strongly optimal error algorithm.

Remark 7.1 We stress that this property of the minimal residual algorithm holds for *every* orthogonally invariant class of matrices. We illustrate this point by the following example. Suppose that one knows *all* eigenvalues of $\bar{A}$ and asks what algorithm should be used for the approximate solution of $\bar{A}x = b$? The surprising answer is that we still should use the minimal residual algorithm (although the *mr* algorithm does not make use of the eigenvalues of $\bar{A}$). To explain this, define

$$F = \{A \colon A = Q\bar{A}Q^T,\ Q \text{ orthogonal}\}.$$

Then all matrices in F have the same eigenvalues as $\bar{A}$. Since F is orthogonally invariant, (7.8) guarantees that the *mr* algorithm will perform at most one step more than is necessary.

This shows that the knowledge of matrix eigenvalues does not help to find the approximate solution of a linear system. ■

For some specific orthogonally invariant classes F, the precise value of $r(N_k)$ or $e(\phi_k^{mr})$ is known. For instance, let

$$F = F_1 = \{A \colon A = A^T > 0,\ \operatorname{cond}(A) \le M\} \tag{7.9}$$

be the class of symmetric and positive definite matrices with condition number $\operatorname{cond}(A) = \| A \| \, \| A^{-1} \|$ bounded by a given constant M, $M \ge 1$. Then, for $k < n$,

$$e(\phi_k^{mr}) = 2q^k(1 + q^{2k}), \qquad q = \frac{\sqrt{M} - 1}{\sqrt{M} + 1} \tag{7.10}$$

and

$$k(F_1) = a_1 + \min\left(n,\ \left\lfloor \ln \frac{1 + \sqrt{1 - \epsilon^2}}{\epsilon} \Big/ \ln \frac{\sqrt{M} + 1}{\sqrt{M} - 1} \right\rfloor + 1\right), \tag{7.11}$$

where $a_1 = 0$ or $a_1 = -1$. For large M and small ϵ, (7.11) can be rewritten as

$$k(F_1) = a_1 + \min\left(n, \frac{\sqrt{M}}{2} \ln \frac{2}{\epsilon} [1 + o(1)]\right). \tag{7.12}$$

Let

$$F = F_2 = \{A : A = A^T, \text{cond}(A) \leq M\} \tag{7.13}$$

be the class of symmetric (not necessarily positive definite) matrices with condition number bounded by M. Then

$$e(\phi_k^{mr}) = \frac{2q^{\lfloor k/2 \rfloor}}{1 + q^{2\lfloor k/2 \rfloor}}, \qquad q = \frac{M-1}{M+1}, \tag{7.14}$$

and

$$k(F_2) = a_2 + \min\left(n, 2 \left\lfloor \ln \frac{1 + \sqrt{1 - \epsilon^2}}{\epsilon} / \ln \frac{M+1}{M-1} \right\rfloor + 2\right), \tag{7.15}$$

where $a_2 = 0$ or $a_2 = -1$. For large M and small ϵ, (7.15) can be rewritten as

$$k(F_2) = a_2 + \min(n, M \ln \frac{2}{\epsilon} [1 + o(1)]). \tag{7.16}$$

Comparing with (7.12), we see how the lack of positive definiteness increases the optimal class index. If n is sufficiently large, i.e., if the minimum in (7.15) is attained for the second argument, we have

$$\frac{k(F_2)}{k(F_1)} = 2\sqrt{M}[1 + o(1)]. \tag{7.17}$$

We now discuss the complexity of this problem. Let the set P of primitive operations consist of matrix-vector multiplications with complexity cn, as well as arithmetic operations and comparisons with unit complexity. Note that if the matrices A are sparse, then c is proportional to unity rather than to n. The minimal residual algorithm ϕ_k^{mr} for the two classes F_1 and F_2 can be implemented with combinatory complexity proportional to nk. From this, (7.7), and (7.8) we conclude that the minimal residual algorithm is a nearly optimal complexity algorithm for the class $\Psi = \{N_1, N_2, \ldots, N_k\}$,

$$\text{comp}(\Psi, \epsilon) \simeq \text{comp}(\phi_{1+k(F_i)}^{mr}) \simeq cnk(F_i), \tag{7.18}$$

where $i = 1, 2$ and $k(F_i)$ is given by (7.11) or (7.15).

Detailed analyses and further results for different orthogonally invariant classes F and various definitions of ϵ-approximation may be found in Traub and Woźniakowski [80b].

(ii) Fixed Precision

In this subsection we deal with approximate information operators for the solution of linear systems. We show that the major concepts of round-off error analysis (such as condition number, numerically stable algorithm) are intimately related to the radius of approximate information. We assume here a model of computation based on floating-point arithmetic with fixed precision. Variable precision will be discussed in the next subsection.

Let $G = \mathbb{R}^n$ with the l_2 norm $\| x \| = \sqrt{(x,x)}$. Let b be a given $n \times 1$ real vector such that $\| b \| = 1$. Let

$$F = \{A\colon A \text{ is a } n \times n \text{ real matrix, } \operatorname{cond}(A) \le M\}. \tag{7.19}$$

Define the solution operator as

$$S(A,\epsilon) = \{g \in G\colon \| g - A^{-1}b \| \le \epsilon \| g \|\}. \tag{7.20}$$

Note that an ϵ-approximation is a vector that approximates the exact solution with relative error not exceeding ϵ. Thus, we consider a different approximation criterion than in subsection (i). Let $N(A) = A$ be complete information. Define the information error operator as

$$E(A,\rho) = \{B\colon \| B - A \| \le \rho \| B \|\}. \tag{7.21}$$

We find the radius $r(N_{\rho,E})$ for small ρ. Let N_ρ be an approximate information operator and let $C = N_\rho(A)$. We assume that

$$\frac{\rho M}{1 - \rho} < 1, \tag{7.22}$$

which guarantees that C is nonsingular. Observe that

$$V(N_\rho,A) = \{\tilde{A} \in F\colon \| C - \tilde{A} \| \le \rho \| C \|\}.$$

Let $Cx_1 = b$ and $\tilde{A}x_2 = b$. Then it is easy to show that

$$\| x_1 - x_2 \| \le \rho \operatorname{cond}(C)(1 + O(\rho)) \| x_1 \|. \tag{7.23}$$

Since (7.23) is sharp, we conclude that

$$r(N_\rho,A) = \rho \operatorname{cond}(N_\rho(A))(1+ O(\rho)) = \rho \operatorname{cond}(A)(1 + O(\rho)). \tag{7.24}$$

From (7.24) and (7.19) we have

$$r(N_{\rho,E}) = \rho M(1 + O(\rho)). \tag{7.25}$$

Thus, we can find an ϵ-approximation iff $\rho M(1 + O(\rho)) \leq \epsilon$. Note that the radius of the worst case approximate information depends linearly on the condition number bound.

Define the algorithm

$$\phi(N_\rho(A)) = N_\rho(A)^{-1}b. \tag{7.26}$$

One can show that ϕ is a nearly strongly optimal error algorithm for N_ρ as well as a nearly strongly optimal algorithm for the worst case, and

$$e(\phi) = r(N_{\rho,E})(1 + O(\rho)). \tag{7.27}$$

We now derive the ϵ-complexity of this problem. Define the set P of primitive operations as the set of arithmetic operations performed in floating-point arithmetic fl_ρ with relative accuracy ρ. That is, for any real floating-point numbers a and b we have

$$fl_\rho(a \mathbin{\square} b) = (a \mathbin{\square} b)(1 + \eta), \tag{7.28}$$

where $\eta = \eta(a,b,\square)$, $|\eta| \leq \rho$, and $\square$ represents one of the four arithmetic operations (we ignore underflow and overflow). We assume that all primitive operations have unit complexity. Note that the algorithm (7.26) is not permissible since it requires the *exact* solution of linear equations. However, if we use a numerically stable algorithm ϕ for the solution of $N_\rho(A)x = b$, then ϕ produces in fl_ρ a vector y such that

$$\| y - N_\rho(A)^{-1}b \| \leq c\rho \operatorname{cond}(N_\rho(A))(1 + O(\rho)) \| y \|,$$

where $c = c(\phi,n)$ depends only on the algorithm used and the problem size. For most numerically stable algorithms, especially if augmented with iterative refinement, one can guarantee that $c(\phi,n)$ is a low degree polynomial in n. This implies that *any numerically stable algorithm is a nearly strongly optimal error algorithm.*

What is the complexity of the linear equation problem? (Recall that complexity equals 0-complexity.) For $\rho = 0$, i.e., if exact arithmetic is used, the complexity is proportional to n^{β}, where the best β known is about 2.49. The precise complexity is unknown.

Suppose now that $\rho > 0$ and the floating-point arithmetic fl_{ρ} is used. What is the minimal complexity of a numerically stable algorithm, i.e., an algorithm that yields a result with relative error of order ρM? The known algorithms have complexity proportional to n^3. It is open if an algorithm with complexity $O(N^{\beta})$, $\beta < 3$, can be numerically stable.

Miller [75] defined the concept of strongly stable algorithm for the matrix multiplication problem and proved that any such algorithm has complexity proportional to n^3. Since the matrix multiplication and linear equation problems are related, it is plausible to conjecture that the complexity of a numerically stable algorithm for the solution of linear equations in floating-point arithmetic is proportional to n^3.

(iii) Variable Precision

Let F, G, A, and N be defined as in subsection (ii). In this subsection we consider variable-precision floating-point arithmetic with the complexity of primitive operations depending on the mantissa length. That is, let P be the set of primitive operations defined as floating-point arithmetic operations on t-binary digit mantissas, where *every* finite integer value of t is permissible. Let $m(t)$ and $a(t)$ denote the cost of multiplying and adding two numbers in floating-point arithmetic with t-binary digit mantissa. Note that the algorithms for multiplying and adding t-binary digit mantissa numbers need not be optimal. We assume that the costs $m(t)$ and $a(t)$ are defined by bit operations and therefore we have

$$\lim_{t\to\infty} m(t) - \lim_{t\to\infty} a(t) = +\infty.$$

We can assume that $a(t)$ is a linear function of t. The minimal value of $m(t)$ is unknown. The best algorithm known for multiplying two t-digit numbers is due to Schönhage and Strassen [71] with $m(t) = \Theta(t \log t \log \log t)$.

Recall that $R(N,\epsilon)$ is the class of permissible algorithms whose errors do not exceed ϵ. Note that known numerically stable algorithms ϕ are permissible. Since the condition number of any matrix A from F is bounded by M, a numerically stable algorithm ϕ provides an ϵ-approximation whenever the computations are performed with sufficiently large t. Thus $R(N,\epsilon)$ is nonempty and the ϵ-complexity comp(N,ϵ) is finite. What is the ϵ-complexity and is it proportional to n^3?

Recently, Kiełbasiński [81] proposed an algorithm ϕ based on iterative refinement implemented in variable precision such that an ϵ-approximation is computed with complexity given by

$$\text{comp}(\phi) = \Theta\left(n^2 m(c + \tau)\left[n\left(\frac{c}{c+\tau}\right)^{\sigma} + \frac{1-\nu^{1-\sigma}}{1-2^{1-\sigma}}\right]\right), \tag{7.29}$$

where $c = \log M$, $\tau = \log(1/\epsilon)$, $\nu = \min(n, 2\tau/c)$, and $m(t) = \Theta(t^\sigma)$.

Thus, if $\sigma > 1$, i.e., we are not using the optimal multiplication algorithm, then, for small ϵ, (7.29) becomes

$$\mathrm{comp}(\phi) = \Theta(n^2 m[\log(1/\epsilon)]). \tag{7.30}$$

This means that the complexity of Kiełbasiński's algorithm depends linearly on n^2 and linearly on the cost of multiplication in the precision determined by ϵ. It is very likely that the ϵ-complexity is also of the form (7.30), and therefore Kiełbasiński's algorithm is a nearly optimal complexity algorithm.

If $\sigma = 1$, i.e., we are using an optimal multiplication algorithm, then, for small ϵ, (7.29) becomes

$$\mathrm{comp}(\phi) = \Theta(n^2 m[\log(1/\epsilon)] \log n). \tag{7.31}$$

Note that it depends on $n^2 \log n$ instead of n^3. In this case it is also plausible to conjecture that ϕ is close to an optimal complexity algorithm.

The same problem in variable precision and in the parallel model of computation was recently studied by Bojańczyk [80]. He presented an algorithm ϕ based on Newton iteration that uses $\Theta(n^3)$ processors and computes an ϵ-approximation with complexity given by

$$\mathrm{comp}(\phi) = \Theta(m(c + \tau) + a(c + \tau)\log n), \tag{7.32}$$

where $\tau = \log(1/\epsilon)$ and $c = \log M$. If $a(t) = o(m(t))$ for large t, (7.32) can be rewritten for small ϵ as

$$\mathrm{comp}(\phi) = \Theta(m[\log(1/\epsilon)]). \tag{7.33}$$

Thus, the complexity of Bojańczyk's algorithm is independent of the size n and is roughly equal to one multiplication performed in the precision determined by ϵ. It is plausible to conjecture that the ϵ-complexity is also of the form (7.32) or (7.33), respectively.

These two examples of Kiełbasiński and Bojańczyk's algorithms illustrate how the ϵ-complexity of the linear equations problem depends on the model of computation (see also Traub and Woźniakowski [82] where the dependence of the complexity of linear programming on the model of computation is discussed).

6.8 FURTHER APPLICATIONS

Numerous additional application areas fit our general framework. In this section we select three more examples. They correspond to finding an information operator whose radius is zero.

(i) Database Security

Let $f = [f_1,f_2,\dots,f_m]$ be a sequence of positive integers f_i. As in Dobkin, Jones, and Lipton [79], we call f a database. Then m denotes the number of objects in the database. One wishes to permit access to statistical data while protecting the value of any one f_i. Assume that queries about the average of any k-element subset of the database are allowed (permissible) and no two queries may overlap in more than one position. The database is compromised if there exists a sequence of allowed queries that can determine the value of some number f_i of the database.

Is it possible to compromise the database with such queries? This and more general problems of database security are addressed by Dobkin, Jones, and Lipton [79]. Related material may be found in DeMillo, Dobkin, and Lipton [78], E. Denning, P. Denning, and Schwartz [79].

We now show how this problem can be formulated and solved in our setting. Let $\mathbb{N}$ be the set of positive integers. Define

$$F = \mathbb{N}^m = \{f: f = [f_1,f_2,\dots,f_m],\ f_i \in \mathbb{N}\}, \qquad G = \mathbb{N}.$$

For $f \in F$ we set

$$S(f,\epsilon) = \{f_1,f_2,\dots,f_m\}, \qquad \forall\epsilon \geq 0.$$

Thus $x \in S(f,\epsilon)$ means that $x = f_i$ for some $i \in [1,m]$. This is equivalent to compromising the database.

We define permissible information operators as follows. Let A be a subset of $\{1,2,\dots,m\}$ consisting of k elements, $k \leq m$. Let

$$U_A(f) = \frac{1}{k}\sum_{i\in A} f_i. \tag{8.1}$$

Thus $U_A(f)$ is the average of the k-element subset of the database. Finally,

$$N_n(f) = [U_{A_1}(f),\ U_{A_2}(f),\dots,\ U_{A_n}(f)] \tag{8.2}$$

is a permissible information operator iff $A_1,A_2,\dots,A_n$ are k-element subsets of $\{1,2,\dots,m\}$ such that $A_i \cap A_j$ has at most one element for $i \neq j$.

We can compromise the database if the radius $r(N_n) = 0$. Thus we ask: how to define the sets $A_1,A_2,\dots,A_n$ to make the radius of (8.2) equal to zero.

The solution of this problem is obtained by Dobkin, Jones, and Lipton [79]. Assume that $n = 2k - 1$ and $k^2 - k + 1 \leq m$. Define

$$\begin{aligned} A_i &= \{k(i-1)+1,\ k(i-1)+2,\dots,k(i-1)+k\}, && i = 1,2,\dots,k-1, \\ A_{k-1+i} &= \{i,i+k,i+2k,\dots,i+(k-2)k,\ k^2-k+1\}, && i = 1,2,\dots,k. \end{aligned} \tag{8.3}$$

Note that each A_i has k elements and $A_i \cap A_j$ is empty or has one element for all $i \neq j$. Thus N_n defined by $A_1, A_2, \ldots, A_n$ is permissible. Define the algorithm

$$\phi(N_n(f)) = \sum_{i=1}^{k} U_{A_{k-1+i}}(f) - \sum_{i=1}^{k-1} U_{A_i}(f).$$

From (8.1) and (8.3) we get

$$\phi(\mathrm{N}_n(f)) = \frac{1}{k}\sum_{i=1}^{k}\left(\sum_{j=1}^{k-1} f_{i+k(j-1)} + f_{k^2-k+1}\right) - \frac{1}{k}\sum_{i=1}^{k-1}\sum_{j=1}^{k} f_{k(i-1)+j}$$

$$= f_{k^2-k+1}.$$

Hence $\phi(N_n(f)) \in S(f,0)$ and $e(\phi) = r(N_n) = 0$. This proves that the database is compromised by queries based on the subsets A_i of (8.3). This means that knowing averages of k elements of the database one can find the value of a single f_i. Note that this holds no matter how large k is, provided that $k^2 - k + 1 \leq m$. Since $n = 2k - 1$, the database is compromised by information of rather low cardinality.

A similar result holds if queries are defined as medians, i.e., $U_A(f) = f_p$, where f_p is the median value of $\{f_i : i \in A\}$ or if $U_A(f) = f_i$ for some (even randomly chosen) i from A.

If one wishes to secure a database while still permitting access to statistical data, a different approach must be taken. One approach is suggested by Traub, Woźniakowski, and Yemini [81].

(ii) Boolean Functions

As we mentioned in Example 3.1 of Chapter 1, we are often interested in finding the minimal amount of information needed for an exact solution. This corresponds to finding the cardinality number since this is the smallest cardinality of an information operator whose radius is equal to zero. We illustrate this by the following problem.

As in Rivest and Vuillemin [76], let $P:\{0,1\}^d \to \{0,1\}$ be a Boolean function. We compute $P(f)$, $f = [f_1, f_2, \ldots, f_d]$ by "decision-tree" algorithms which successively examine the various components of f. Rivest and Vuillemin ask what is the minimal number of components of f that must be known to compute $P(f)$.

We formulate this question in our setting. Let

$$F = \{0,1\}^d, \qquad G = \{0,1\}$$

and

$$S(f,\epsilon) = \{P(f)\}, \qquad \forall\ \epsilon \geq 0.$$

Let $H_1 = \{0,1\}$ and $U_i(f) = f_i$. The class $\Psi(H_1)$ of primitive information operators is defined as

$$\Psi(H_1) = \{U_1, U_2, \ldots, U_d\}.$$

The information operator N_n, which consists of n primitive information operators, has the form

$$N_n(f) = [f_{\sigma(1)}, f_{\sigma(2)}, \ldots, f_{\sigma(n)}],$$

where σ is a permutation of $\{1,2,\ldots,d\}$. Note that N_n is adaptive since $\sigma(i)$ may depend on $f_{\sigma(1)}, f_{\sigma(2)}, \ldots, f_{\sigma(i-1)}$.

Consider the class of idealized algorithms, $R = \Phi(N)$. Then the cardinality number $m^a(0)$ is the smallest n such that there exists N_n with $r(N_n) = 0$. Obviously, $m^a(0) \leq d$.

Rivest and Vuillemin show that for most Boolean functions P, $m^a(0) = d$. In particular, they proved that if P is transitive, $P(0) \neq P(1)$, and d is a prime power, then $m^a(0) = d$. This means that we have to know all components of f in order to compute $P(f)$ for all f from F.

(iii) Linear Equations

This is the second example of this section where we are interested in the cardinality number. Consider the system of linear equations $Ax = b$, where A is a $m \times m$ real nonsingular matrix and b is a $m \times 1$ real vector. As in Rabin [72], assume that we can compute the inner products (called *tests*) $a_i^T v - b_i$, where a_i is the ith row of A and b_i is the ith component of b, for an arbitrary $m \times 1$ vector v. Rabin asks: What is the minimal number of tests required to solve $Ax = b$ exactly?

We formulate this question in our setting. Let

$$F = \{f = (A,b)\colon A \text{ is a } m \times m \text{ nonsingular matrix, } b \text{ is a } m \times 1 \text{ vector}\},$$

$$G = \mathbb{R}^m,$$

and

$$S(f,\epsilon) = \{A^{-1}b\}, \qquad \forall\ \epsilon \geq 0.$$

Let $H_1 = \mathbb{R}$ and $U(f) = U(f;i,v) = a_i^T v - b_i$ be a primitive information operator. Let

$$\Psi(H_1) = \{U(\cdot;i,v): v \in \mathbb{R}^m, \; i \in [1,m]\}$$

be the class of primitive information operators. Then the information operator N_n consisting of n primitives has the form

$$N_n(f) = [U(f;i_1,v_1),..., \; U(f;i_n,v_n)].$$

The information N_n is adaptive since i_j and v_j may depend on $U(f;i_1,v_1),...,$ $U(f;i_{j-1},v_{j-1})$.

What is the smallest n for which there exists N_n with $r(N_n) = 0$? It is obvious that if $n = m(m + 1)$, then we can recover all elements of A and b, and then solve $Ax = b$. Thus for the class of idealized algorithms,

$$m^a(0) \leq m(m + 1).$$

Is this bound sharp? The answer is no since Rabin exhibits an algorithm similar to Gaussian elimination that uses $n = m(m + 3)/2$ tests. Thus

$$m^a(0) \leq m(m + 3)/2.$$

Is this improved bound sharp? This question is open. However, assuming that the computation of the information operators N_n involves only rational operations, Rabin proves the sharpness of the last bound.

6.9 AVERAGE CASE MODELS: INFORMATION THEORY AND DECISION THEORY

To demonstrate the generality of our setting, we indicate how problems from information and decision theory may be formulated in our framework. Since they would lose much of their characteristic flavor in our worst case model, we shall deviate from the main stream of this monograph and introduce an average case model. This section foreshadows some of our future work in which we study various probabilistic models. We expect to present a far more thorough examination of information and decision theory in that work.

In our discussion of information theory, we shall introduce the concept of average cardinality number and prove that it differs by at most unity from Shannon entropy. We stress that average cardinality number is an important concept for the average case model and will be studied in future work.

Decision theory is one of many areas in which the concept of state is important. The set of states plays a role here analogous to the role of, say, the set of integrands in integration. In our abstract theory, state and integrand are both examples of problem elements.

We first discuss an example drawn from information theory, which is followed by an example from decision theory.

(i) Information Theory

We showed in GTOA (pp. 159–162) and in subsection (iii) of Section 5.4 of Chapter 5 how the ϵ-cardinality number and ϵ-entropy are related in the worst case model. Here we show that for an average case model, Shannon entropy and average cardinality number differ by at most unity for an example from information theory.

Remark 9.1 More accurately, information theory might be called a mathematical theory of communication, which is how Shannon titled his seminal paper (Shannon [48]). Hamming [80, p.2] defines it as "an abstract mathematical theory of the representation of some undefined source symbols in terms of a fixed alphabet." ■

Let $F = G = \{f_1, f_2, \ldots, f_n\}$ for some symbols f_i, $n \geq 2$. Let $S(f,\epsilon) = \{f\}$, $f \in F$. We want to transmit f_i by sending binary digits in such a way that the average number of binary digits sent is minimized. That is, let

$$N(f) = [g_1(f), \ldots, g_k(f)], \tag{9.1}$$

where $g_i(f) \in \{0,1\}$ and $k = k(f)$. Thus $N(f)$ is the encoding of f. If N is one-to-one, then the algorithm

$$\phi(N(f)) = f \tag{9.2}$$

is well defined and represents the decoding of f. Clearly $e(\phi) = r(N) = 0$ iff N is one-to-one.

Let N be one-to-one. Since we have n elements, $k(f) \geq \log n$ for some f. Thus, if the cardinality of N is defined by the worst case, i.e.,

$$\text{card}(N) = \max_{f \in F} k(f),$$

then $\text{card}(N) \geq \log n$.

Let Ψ be the class of one-to-one information operators of the form (9.1). Then it is easy to check that the (worst case) cardinality number $m(0)$,

$$m(0) = \min\{\text{card}(N) : N \in \Psi, N \text{ is one-to-one}\},$$

is given by $m(0) = \lceil \log n \rceil$.

We now discuss the average case. Assume that f_i occurs with probability $\mathrm{pr}(f_i)$. Define the *average cardinality of N* as

$$\mathrm{card}^{\mathrm{avg}}(N) = \sum_{i=1}^{n} \mathrm{pr}(f_i)\, k(f_i). \tag{9.3}$$

Let $m^{\mathrm{avg}}(0) = \min\{\mathrm{card}^{\mathrm{avg}}(N) : N \in \Psi,\ N \text{ is one-to-one}\}$ be the *average cardinality number*. Then, in our notation, Shannon's noiseless coding theorem gives the average cardinality number as

$$m^{\mathrm{avg}}(0) = a - \sum_{i=1}^{n} \mathrm{pr}(f_i) \log \mathrm{pr}(f_i), \tag{9.4}$$

where $a \in [0,1]$. Thus the entropy differs from the average cardinality number by at most unity. The average cardinality number may be significantly smaller than the worst case cardinality number.

We now show that the worst case and average cardinality numbers are equal to the worst case and average complexities for a particular set of primitive operations. Let the set P of primitive operation be defined as $P = \{p_1, p_2, p_3\}$, where $p_i : F \to \{0,1\}$, $p_1(f) = 0$, $p_2(f) = 1$, $\forall f \in F$, with $\mathrm{comp}(p_1) = \mathrm{comp}(p_2) = 1$, and p_3 is any mapping such that $p_3 : [0,1]^m \to F$ for some m, $\mathrm{comp}(p_3) = 0$.

The primitives p_1 and p_2 serve to encode f, and $N(f)$ defined by (9.1) is formed by $k(f)$ of them, yielding

$$\mathrm{comp}(N(f)) = k(f). \tag{9.5}$$

The primitive p_3 is a decoding device with zero complexity. If N is one-to-one, then the algorithm ϕ defined by (9.2) is permissible and

$$\mathrm{comp}(\phi(N(f)) = k(f).$$

Since $k(f) \geq \lceil \log n \rceil$ for some f, the worst case complexity of an algorithm ϕ defined as in (3.1) of Chapter 5 satisfies

$$\mathrm{comp}(\phi) \geq \lceil \log n \rceil .$$

Since the combinatory complexity is zero, this yields

$$\mathrm{comp}(N) \geq \lceil \log n \rceil$$

whenever N is one-to-one. Then it is easy to check that for the class Ψ we have

$$\mathrm{comp}(\Psi) = m(0) = \lceil \log n \rceil . \tag{9.6}$$

Note that comp(Ψ) is the minimal number of binary bits required to encode n elements for a worst case. Thus, (9.6) establishes equality between the worst case complexity and the worst case cardinality number.

We now proceed to the average case. Define the average complexity of the algorithm (9.2) using the information (9.1) as

$$\text{comp}^{\text{avg}}(N) = \text{comp}^{\text{avg}}(\phi) = \sum_{i=1}^{n} \text{pr}(f_i)\ \text{comp}(\phi(N(\text{f}_i))).$$

Let

$$\text{comp}^{\text{avg}}(\Psi) = \min_{N \in \Psi} \text{comp}^{\text{avg}}(N)$$

be the average complexity. Then (9.5), (9.3), and Shannon's noiseless coding theorem yield

$$\text{comp}^{\text{avg}}(\Psi) = m^{\text{avg}}(0) = a - \sum_{i=1}^{n} \text{pr}(f_i) \log \text{pr}(f_i).$$

Thus, the average complexity equals the average cardinality number and they both differ from the entropy by at most unity.

Some related material on the average cardinality number and the average complexity for approximate information may be found, for instance, in Pearl [76].

(ii) Decision Theory

We show how an example from decision theory can be formulated in our setting. The example is drawn from Marschak and Radner [72]. We have a triplet (F, G, w), where F and G are sets and w, $w\colon F \times G \to \mathbb{R}$, is a given function. The set F is called the *set of states*, and an element f of F is called *nature, environment,* or the *external world.* The set G is called the *set of conceivable actions.* The function w is called a *gross payoff function* and $w(f, g)$ is the *payoff* to the decision maker when an action g is taken and f is the state of environment. For simplicity assume that both F and G are finite. Let pr, pr: $F \to [0,1]$, be a probability measure on $F = \{f_1, f_2, \ldots, f_n\}$.

Let N, $N\colon F \to H$, be an information operator (called the *information structure*) and let ϕ, $\phi : N(F) \to G$, be an algorithm (called a *decision function*). Then

$$\Omega(\phi) \overset{df}{=} \sum_{i=1}^{n} w(f_i, \phi(N(f_i))\ \text{pr}(f_i) \tag{9.7}$$

is the expected payoff. The best decision function ϕ^* is defined as a decision function which uses N and maximizes the expected payoff,i.e.,

$$\Omega(\phi^*) = \sup_{\phi \in \Phi(N)} \Omega(\phi). \tag{9.8}$$

We now show how the problem of the best decision function can be formulated in our setting. Let $w(f) = \max_g w(f,g)$ and define the solution operator as

$$S(f,\epsilon) = \{g \in G\colon w(f) - w(f,g) \le \epsilon\}.$$

Let ϕ be an algorithm that uses N. Then

$$\inf\{\delta\colon \phi(N(f)) \in S(f,\delta)\} = w(f) - w(f,\phi(N(f))$$

and the error of ϕ (defined by the worst case) is given by

$$\begin{aligned} e(\phi,N) &\overset{df}{=} \inf\{\delta\colon \phi(N(f)) \in S(f,\delta),\ \forall f \in F\} \\ &= \max_{f \in F} \{w(f) - w(f,\phi(N(f)))\}. \end{aligned} \tag{9.9}$$

We define the average error of ϕ as

$$\begin{aligned} e^{\text{avg}}(\phi,N) &= \sum_{i=1}^{n} \{w(f_i) - w(f_i,\phi(N(f_i)))\}\ \text{pr}(f_i) \\ &= \sum_{i=1}^{n} w(f_i)\ \text{pr}(f_i) - \Omega(\phi). \end{aligned} \tag{9.10}$$

Let the average radius of information be the infimum over average errors of algorithms that use N, i.e.,

$$r^{\text{avg}}(N) = \inf_{\phi \in \Phi(N)} e^{\text{avg}}(\phi,N). \tag{9.11}$$

Then

$$r^{\text{avg}}(N) = \sum_{i=1}^{n} w(f_i)\ \text{pr}(f_i) - \sup_{\phi \in \Phi(N)} \Omega(\phi). \tag{9.12}$$

The best decision function is an algorithm whose average error equals the average radius of information.

It was noted by Marschak and Radner [72, p. 52] that the best decision function maximizes the conditional expected payoff for all $N(f_i)$. In our terminology this corresponds to the following: every optimal error algorithm is also strongly optimal. In future work, we shall show that this result is true in general for the average case model. This means that the difference between optimal error algorithms and strongly optimal algorithms, which occurs in the worst case model, disappears in the average case model.

APPENDIX A

INTERPOLATORY ALGORITHMS

Interpolatory algorithms play an important role in the normed linear space setting. In that setting the class of interpolatory algorithms is defined as follows. Choose an element $\tilde{f}$ indistinguishable from f under N. That is, choose $\tilde{f} \in F$ such that $N(\tilde{f}) = N(f)$ and such that $S(\tilde{f})$ is "simple" to calculate. Then we define $\phi(N(f)) = S(\tilde{f})$.

Interpolatory algorithms enjoy three desirable properties in the normed linear space setting:

(i) $e(\phi^I, N, f) \le 2r(N, f)$, $\forall f \in F$; therefore ϕ^I is almost strongly optimal;

(ii) Interpolatory algorithms are conceptually simple;

(iii) An interpolatory algorithm is often far easier to construct than a central algorithm.

In this appendix we shall give a natural generalization of interpolatory algorithms. In general, interpolatory algorithms are not almost strongly optimal. Indeed the ratio of the local error of an interpolatory algorithm to the local radius may be unbounded. We shall characterize for which S interpolatory algorithms are guaranteed to enjoy good optimality properties.

Let N_ρ be an approximate information operator and let $R = R(N_\rho)$ be a class of realizable algorithms. Knowing $N_\rho(f)$, we choose an element $\tilde{f}$ from $V(N_\rho, f)$ such that $S(\tilde{f}, \delta) \cap Q(N_\rho(f))$ is nonempty for some nonnegative δ. We define $\phi^I(N_\rho(f))$ as any element of $S(\tilde{f}, \delta) \cap Q(N_\rho(f))$. More precisely, we have the following definition.

An algorithm ϕ^I is called *δ-interpolatory* iff for every $f \in F$ there exists $\tilde{f} \in V(N_\rho, f)$ such that

$$\phi^I(N_\rho(f)) \in S(\tilde{f}, \delta) \cap Q(N_\rho(f)). \tag{A.1}$$

An algorithm that is 0-interpolatory is simply called *interpolatory*. Obviously, δ_1-interpolatory is also δ_2-interpolatory whenever $\delta_1 \le \delta_2$. We illustrate an δ-interpolatory algorithm by an example.

Example: Normed Linear Space (continued) Let $R = \Phi(N)$ and $\rho = 0$. Let ϕ^I be δ-interpolatory. This means that $\phi^I(N(f)) = S(\tilde{f})$ for $\delta = 0$ or $\| \phi^I(N(f)) - S(\tilde{f}) \| < \delta$ for $\delta > 0$, where $\tilde{f} \in V(N,f)$. Thus for $\delta = 0$, (A.1) coincides with the definition of interpolatory algorithms in normed linear spaces (see GTOA, p. 12).

We estimate the local error of ϕ^I. We have

$$e(\phi^I,N,f) = \sup_{g \in V(N,f)} \| S(g) - \phi^I(N(f)) \| \leq \| \phi^I(N(f)) - S(\tilde{f}) \|$$

$$+ \sup_{g \in V(N,f)} \| S(\tilde{f}) - S(g) \| \leq \delta + \operatorname{diam}(U(f)),$$

where $\operatorname{diam}(U(f))$ is the diameter of $U(f) = S(V(N,f))$. Since the diameter of any set is less than twice the radius, we conclude that

$$e(\phi^I,N,f) \leq \delta + 2\, r(N,f).$$

This shows that for δ small relative to $r(N,f)$, any δ-interpolatory algorithm has local error comparable to the minimal one. ∎

Let ϕ be a realizable algorithm with finite error $e = e(\phi,N_\rho)$. Then

$$\phi(N_\rho(f)) \in A(N_\rho,f,e + \eta) \cap Q(N_\rho(f)), \qquad \forall \eta > 0, \qquad \forall f \in F.$$

Let $\delta > e$. Setting $\eta = \delta - e$, we conclude that

$$\phi(N_\rho(f)) \in S(\tilde{f},\delta) \cap Q(N_\rho(f)), \qquad \forall \tilde{f} \in V(N_\rho,f).$$

This proves

COROLLARY A.1 An algorithm ϕ from R is δ-interpolatory for any $\delta > e(\phi,N_\rho)$. In particular, a central algorithm from R is δ-interpolatory for any $\delta > r(R,N_\rho)$. ∎

For $R = \Phi(N)$, we know that a central algorithm is strongly optimal. Then Corollary A.1 states that there exists an δ-interpolatory algorithm with strong optimal error properties whenever $\delta > r(N_\rho)$. Is this true for $\delta \leq r(N_\rho)$? We shall prove that, in general, this is *not* the case. That is, there exists a problem such that all δ-interpolatory algorithms have infinite error provided δ is no larger than the radius of information. More precisely, we prove

THEOREM A.1 For any number $d \in [1,+\infty]$ there exist S and N such that for any E and N_ρ we have

(i) $r(N_\rho,f) = r(N_\rho) = 1, \qquad \forall f \in F,$

(ii) $e(\phi^I,N_\rho,f) = d, \qquad \forall f \in F,$

for any δ-interpolatory algorithm ϕ^I with $\delta \le r(N_\rho) = 1$. ■

PROOF Let $F = \{0,2\}$ and $G = \mathbb{R}$. Let

$$\mu(a,b) = \begin{cases} |a - b| & \text{if } |a - b| \le 1, \\ d & \text{otherwise.} \end{cases}$$

Define the solution operator as

$$S(f,0) = \{f\}, \qquad S(f,\epsilon) = \{x: \mu(f,x) < \epsilon\}, \epsilon > 0.$$

Thus

$$S(f,\epsilon) = \begin{cases} \{f\} & \text{if } \epsilon = 0, \\ (f-\epsilon,f+\epsilon) & \text{if } 0 < \epsilon \le 1, \\ [f-1,f+1] & \text{if } 1 < \epsilon \le d, \\ (-\infty,+\infty) & \text{if } d < \epsilon < +\infty. \end{cases}$$

Let $N(f) \equiv 0$. Since $N_\rho(f) \in E(N(f),\rho) = E(0,\rho) = E(N(\tilde{f}),\rho)$, $\forall \tilde{f} \in F$, we have $V(N_\rho,f) = F$ and $A(N_\rho,f,\delta) = S(0,\delta) \cap S(2,\delta)$. Further, $A(N_\rho,f,\delta) = \emptyset$ for $\delta \le 1$ and $1 \in A(N_\rho,f,1 + \delta)$, $\forall \delta > 0$. Thus

$$r(N_\rho,f) = \inf\{\delta: A(N_\rho,f,\delta) \ne \emptyset\} = 1, \qquad \forall f \in F,$$

which proves condition (i).

Let ϕ^I be any δ-interpolatory algorithm with $\delta \le 1$. Then $\phi^I(N_\rho(f)) \in S(\tilde{f},\delta)$, where $\tilde{f} = 0$ or $\tilde{f} = 2$. Without loss of generality assume that $\tilde{f} = 0$. Then $S(0,0) = \{0\}$ and $S(0,\delta) = (-\delta,\delta)$ for $\delta \in (0,1]$. Since $S(2,\eta) \subset [1,3]$ for $\eta \le d$, then $S(0,\delta) \cap S(2,\eta) = \emptyset, \forall \eta \le d$. Thus $\phi^I(N_\rho(f)) \notin S(2,\eta)$, $\forall \eta \le d$, which proves that $e(\phi^I,N_\rho,f) \ge d$, $\forall f \in F$. Since $\phi^I(N_\rho(f)) \in \mathbb{R} = S(2,d + \delta)$, $\forall \delta > 0$, then $e(\phi^I,N_\rho,f) = d$, $\forall f \in F$. Hence condition (ii) is proven. ■

Remark A.1 There exists a unique central algorithm for the problem introduced in the proof of Theorem A.1. Indeed, since $r(N_\rho,f) = 1$ and $1 \in A(N_\rho,f,1 + \delta)$, $\forall \delta > 0$, then $\phi^c(N_\rho(f)) = 1$ is the desired algorithm. Note that the central algorithm ϕ^c is *not* δ-interpolatory for any $\delta \le r(N_\rho)$. This proves that the statement in Corollary A.1 is best possible. ■

Theorem A.1 states that as long as $\delta \leq r(N_\rho)$, then a δ-interpolatory algorithm is not, in general, an optimal error algorithm. From Corollary A.1 it follows that the bound $\delta \leq r(N_\rho)$ is sharp for $R = \Phi(N_\rho)$.

It is natural to ask for which problems S all δ-interpolatory algorithms (with $\delta \leq r(R,N_\rho)$) are nearly strongly optimal error algorithms. We know that the class of such S is not empty since this property holds for the normed linear space setting with $R = \Phi(N)$ and $\rho = 0$. Note that for an operator S defined as in the normed linear space problem, $S(f,\delta)$ grows to the space G as δ tends to infinity, and this holds for any f. As we shall see, in the general setting this condition is replaced by one regarding how $S(f,\delta)$ behaves as a function of δ.

Recall that

$$A(N_\rho,f,\eta) = \bigcap_{\tilde{f}\in V(N_\rho,f)} S(\tilde{f},\eta).$$

Let

$$B(N_\rho,f) = \bigcup_{\tilde{f}\in V(N_\rho,f)} S(\tilde{f},0). \tag{A.2}$$

For a given $f \in F$ and a nonnegative δ define

$$\begin{aligned} h(f,\delta) = \sup_{\tilde{f}\in V(N_\rho,f)} & [\inf\{\eta: S(\tilde{f},\delta) \cap Q(N_\rho(f)) \subset A(N_\rho,f,\eta)\} \\ & - \inf\{\eta: S(\tilde{f},0) \cap Q(N_\rho(f) \subset A(N_\rho,f,\eta)\}], \end{aligned} \tag{A.3}$$

$$d(f) = \frac{\inf\{\eta: B(N_\rho,f) \cap Q(N_\rho(f)) \subset A(N_\rho,f,\eta)\}}{\inf\{\eta: \emptyset \neq Q(N_\rho(f)) \cap A(N_\rho,f,\eta)\}}, \tag{A.4}$$

with the convention that $+\infty - \infty = 0$, $\infty/\infty = 0/0 = 1$. Thus $h(f,\delta)$ is roughly the difference between the smallest η_1, such that $S(\tilde{f},\delta) \cap Q(N_\rho(f))$ is a subset of $A(N_\rho,f,\eta_1)$, and the smallest η_2, such that $S(\tilde{f},0) \cap Q(N_\rho(f))$ is a subset of $A(N_\rho,f,\eta_2)$ for the worst $\tilde{f}$ from $V(N_\rho,f)$. Of course, $\eta_1 \geq \eta_2$. Thus $h(f,\delta) \geq 0$ and $h(f,0) = 0$. Note that $d(f)$ is roughly the ratio between the smallest η_3, such that $B(N_\rho,f) \cap Q(N_\rho(f))$ is a subset of $A(N_\rho,f,\eta_3)$, and the smallest η_4, such that $A(N_\rho,f,\eta_4) \cap Q(N_\rho(f))$ is not empty. Note that the smallest η_4 is equal to $r(R,N_\rho,f)$. Of course, $\eta_3 \geq \eta_4$. Thus $d(f) \geq 1$. All the η_i mentioned above measure how fast the set $A(N_\rho,f,\eta)$ grows to contain a given set. We are ready to prove

THEOREM A.2 Let $\Phi^I(\delta)$ be the class of all δ-interpolatory algorithms. Then

$$\max(h(f,\delta),\ d(f)\ r(R,N_\rho,f)) \leq \sup_{\phi\in\Phi^I(\delta)} e(\phi,N_\rho,f) \tag{A.5}$$

$$\leq h(f,\delta) + d(f)\, r(R,N_\rho,f), \qquad \forall f \in F,$$

with the convention that $+\infty \cdot 0 = +\infty$. ■

PROOF First assume that $r(R,N_\rho,f) = +\infty$. Then, by Theorem 3.1 of Chapter 3, $e(\phi,N_\rho,f) = +\infty$, $\forall\phi$, and (A.5) holds trivially. We can therefore assume that $r(R,N_\rho,f) < +\infty$.

We first prove the right-hand side of (A.5). Let ϕ^I be δ-interpolatory. Then

$\phi^I(N_\rho(f)) \in S(\tilde{f},\delta) \cap Q(N_\rho(f))$ for some $\tilde{f} \in V(N_\rho,f)$ and

$$\begin{aligned} e(\phi^I,N_\rho,f) &= \inf\{\eta\colon \phi^I(N_\rho(f)) \in A(N_\rho,f,\eta)\} \\ &\leq \inf\{\eta\colon S(\tilde{f},\delta) \cap Q(N_\rho(f)) \subset A(N_\rho,f,\eta)\} \\ &\leq \sup_{\tilde{f}\in V(N_\rho,f)} \inf\{\eta\colon S(\tilde{f},\delta) \cap Q(N_\rho(f)) \subset A(N_\rho,f,\eta)\} \\ &\quad - \inf\{\eta\colon S(\tilde{f},0) \cap Q(N_\rho(f)) \subset A(N_\rho,f,\eta)\} \\ &\quad + \sup_{\tilde{f}\in V(N_\rho,f)} \inf\{\eta\colon S(\tilde{f},0) \cap Q(N_\rho(f)) \subset A(N_\rho,f,\eta)\} \\ &= h(f,\delta) + \inf\{\eta\colon B(N_\rho,f) \cap Q(N_\rho(f)) \subset A(N_\rho,f,\eta)\} \\ &= h(f,\delta) + d(f)\, r(R,N_\rho,f). \end{aligned}$$

Since ϕ^I is arbitrary, the right-hand side of (A.5) is proven.

We now prove the left-hand side of (A.5). Assume first that $h(f,\delta)$ and $d(f)$ are both finite. Let a be a positive real. Choose f_1 from $V(N_\rho,f)$ such that

$$b - \inf\{\eta\colon S(f_1,0) \cap Q(N_\rho(f)) \subset A(N_\rho,f,\eta)\} \geq h(f,\delta) - \tfrac{a}{2}\}, \quad \text{(A.6)}$$

where $b = \inf\{\eta\colon S(f_1,\delta) \cap Q(N_\rho(f)) \subset A(N_\rho,f,\eta)\}$. Assume that $b = +\infty$. Then

$$\inf\{\eta\colon S(f_1,0) \cap Q(N_\rho(f)) \subset A(N_\rho,f,\eta)\} = +\infty$$

and consequently $d(f) = +\infty$, which is a contradiction. Thus $b < +\infty$. Next observe that

$$b = \sup_{x\in S(f_1,\delta)\cap Q(N_\rho(f))} \inf\{\eta\colon x \in A(N_\rho,f,\eta)\}$$

and choose an element x_1 from $S(f_1,\delta) \cap Q(N_\rho(f))$ such that

$$\inf\{\eta: x_1 \in A(N_\rho,f,\eta)\} \geq b - a/2. \tag{A.7}$$

Define

$$\phi_1(N_\rho(f)) = x_1.$$

Then ϕ_1 is δ-interpolatory and (A.7), (A.6) yield

$$e(\phi_1,N_\rho,f) \geq b - a/2 \geq h(f,\delta) - a. \tag{A.8}$$

Since a is arbitrary, $\sup\{e(\phi^I,N_\rho,f) : \phi^I \in \Phi^I(\delta)\} \geq h(f,\delta)$. Observe next that

$$\begin{aligned} d(f)\, r(R,N_\rho,f) &= \inf\{\eta: B(N_\rho,f) \cap Q(N_\rho(f)) \subset A(N_\rho,f,\eta)\} \\ &= \sup_{\tilde{f}\in V(N_\rho,f)} \inf\{\eta: S(\tilde{f},0) \cap Q(N_\rho(f)) \subset A(N_\rho,f,\eta)\} \end{aligned}$$

is finite. Choose f_2 from $V(N_\rho,f)$ such that

$$\inf\{\eta: S(f_2,0) \cap Q(N_\rho(f)) \subset A(N_\rho,f,\eta)\} \geq d(f)\, r(R,N_\rho,f) - a/2 \tag{A.9}$$

and an element x_2 from $S(f_2,0) \cap Q(N_\rho(f))$ such that

$$\inf\{\eta: x_2 \in A(N_\rho,f,\eta)\} \geq \inf\{\eta: S(f_2,0) \cap Q(N_\rho(f)) \subset A(N_\rho,f,\eta)\} - a/2 \tag{A.10}$$

Then the algorithm $\phi_2(N_\rho(f)) = x_2$ is interpolatory (thus it is also δ-interpolatory) and (A.10), (A.9) yield

$$e(\phi_2,N_\rho,f) \geq d(f)\, r(R,N_\rho,f) - a.$$

Since a is arbitrary, $\sup\{e(\phi^I,N_\rho,f) : \phi^I \in \Phi^I(\delta)\} \geq d(f)\, r(R,N_\rho,f)$, which proves the left-hand side of (A.5).

To finish the proof, assume that $h(f,\delta)$ or $d(f)$ is infinity. If $h(f,\delta) = +\infty$, then for some f_1 from $V(N_\rho,f)$ and some x_1 from $S(f_1,\delta) \cap Q(N_\rho(f))$, $\inf\{\eta: x_1 \in A(N_\rho,f,\eta)\}$ is arbitrarily large. Thus the algorithm $\phi_1(N_\rho(f)) = x_1$ is δ-interpolatory and has arbitrarily large local error. Thus $\sup\{e(\phi^I,N_\rho,f): \phi^I \in \Phi^I(\delta)\} = +\infty$. Similarly, if $d(f) = +\infty$, then for some f_2 from $V(N_\rho,f)$ and x_2 from $S(f_2,0) \cap Q(N_\rho(f))$, $\inf\{\eta: x_2 \in A(N_\rho,f,\eta)\}$ is arbitrarily large. Then the algorithm $\phi_2(N_\rho(f)) = x_2$ is interpolatory and $e(\phi_2,N_\rho,f)$ is also arbitrarily large. Hence $\sup\{e(\phi^I,N_\rho,f) : \phi^I \in \Phi^I(\delta)\} = +\infty$, which finally completes the proof. ■

Theorem A.2 states that the local errors of δ-interpolatory algorithms depend essentially on the sum $h(f,\delta) + d(f)\, r(R,N_\rho,f)$. Note that for $\delta = 0$, $h(f,0) = 0$, which yields

COROLLARY A.2 $\sup_{\phi \in \Phi'(0)} e(\phi, N_\rho, f) = d(f)\, r(R, N_\rho, f), \quad \forall f \in F.$ ■

Note that from Theorem A.1 it follows that $d(f)$ can be an arbitrary number from the interval $[1, +\infty]$. For the problems defined in normed linear spaces with $R = \Phi(N)$ and $\rho = 0$, $d(f) \le 2$. We end this appendix by two examples.

Example: Binary Search (continued) We find $h(f,\delta)$ and $d(f)$ for $R = \Phi(N)$ and $\rho = 0$. We use the notation from Chapter 1. We need to find the smallest η such that $S(\tilde{f},\delta) \subset A(N,f,\eta)$. This is equivalent to

$$\{\tilde{f} - \lfloor \delta m \rfloor, \ldots, \tilde{f} + \lfloor \delta m \rfloor\} \subset \{f_2 - \lfloor \eta m \rfloor, \ldots, f_1 + \lfloor \eta m \rfloor\}$$

Thus

$$\eta = \frac{p - 1 + \lfloor \delta m \rfloor}{m}$$

for the worst $\tilde{f}$, which yields

$$h(f,\delta) = \frac{\lfloor \delta m \rfloor}{m}, \qquad \forall f \in F.$$

Since $B(f) = V(N,f)$, then the smallest η, such that $V(N,f) \subset A(N,f,\eta)$, is given by $\eta = (p-1)/m$. Thus for any $f \in F$,

$$d(f) = \begin{cases} 2\,\dfrac{p-1}{p} & \text{if } p \text{ is even,} \\ 2 & \text{if } p \text{ is odd, } p \ge 3, \\ 0 & \text{if } p = 1. \end{cases}$$

Note that for $p = 2$, $d(f) = 1$, which means that any interpolatory algorithm is central. ■

Example: Integration (continued) We also find $h(f,\delta)$ and $d(f)$ for $R = \Phi(N)$ and $\rho = 0$. We use the notation from Chapter 1. Observe that $S(\tilde{f},\delta) \subset A(N,f,\eta)$ means $[c - \delta, c + \delta] \subset [b - \eta, a + \eta]$, where $c = \int_0^1 \tilde{f}(t)\,dt$. Thus the smallest η is given by

$$\eta = \max(b - c + \delta,\ c - a + \delta).$$

Hence

$$h(f,\delta) = \delta, \qquad \forall f \in F.$$

Since $B(f) = [a,b]$, then the smallest η such that $[a,b] \subset A(N,f,\eta)$ is given by $\eta = b - a$. Thus

$$d(f) = 2, \qquad \forall f \in F. \qquad \blacksquare$$

APPENDIX B

INTERIOR ALGORITHMS

At the beginning of Appendix A we listed three desirable properties that interpolatory algorithms enjoy in the normed linear space setting. The first of these properties is that interpolatory algorithms are almost strongly optimal. We showed that this property does not hold in the general setting.

In this appendix we introduce the concept of interior algorithm and show that interior algorithms are always almost strongly optimal. However, interior algorithms are conceptually more difficult than interpolatory algorithms and are also harder to construct. For simplicity, we confine ourselves to the class of idealized algorithms.

We also introduce the concept of midpoint algorithm. Every midpoint algorithm is strongly optimal.

Interior algorithms will play an important role in the probabilistic models which will be analyzed in future publications.

Since we don't use a norm, we don't have the notion of distance in the general setting. We begin our analysis by introducing two functionals which may be thought of as generalized distances. For a problem S we define two functionals $\text{dist}_G: F \times G \to \mathbb{R}_+$ and $\text{dist}: F \times F \to \mathbb{R}_+$:

$$\text{dist}_G(f,g) = \inf\{\delta: g \in S(f, \delta)\} \tag{B.1}$$

and

$$\begin{aligned}\text{dist}(f_1,f_2) &= \inf_{g \in G} \{\text{dist}_G(f_1,g) + \text{dist}_G(f_2,g)\} \\ &(= \inf\{\delta_1 + \delta_2: S(f_1,\delta_1) \cap S(f_2, \delta_2) \neq \emptyset\}).\end{aligned} \tag{B.2}$$

As noted above, these two functionals can be thought of as generalized distances; $\text{dist}_G(f,g)$ "measures" the distance between elements of F and G, while $\text{dist}(f_1,f_2)$ "measures" the distance between elements of F. Observe that

$$\text{dist}(f_1,f_2) \le 2 \inf\{\delta: S(f_1,\delta) \cap S(f_2,\delta) \neq \emptyset\}. \tag{B.3}$$

Let A be a subset of F. Then

$$S(A,0) = \bigcup_{f \in A} S(f,0)$$

is a subset of G. For brevity we shall write $S(A) \stackrel{df}{=} S(A,0)$. The point m, $m \in G$, is called an *interior point of* $S(A)$ iff there exists a mapping P, $P: A \to A$, such that

$$\operatorname{dist}_G(f,m) + \operatorname{dist}_G(P(f),m) = \operatorname{dist}(f,P(f)), \qquad \forall f \in A. \tag{B.4}$$

If, in addition,

$$\operatorname{dist}_G(f,m) = \operatorname{dist}_G(P(f),m), \qquad \forall f \in A, \tag{B.5}$$

then m is called a *midpoint of* $S(A)$. Note that the concepts of interior point and midpoint depend on the set A as well as on the set $S(A)$.

Let N_ρ be an approximate information operator and suppose that $S(V(N_\rho,f)$ has an interior point $m(f)$, $\forall f \in F$. Then an algorithm ϕ,

$$\phi(N_\rho(f)) = m(f), \qquad \forall f \in F, \tag{B.6}$$

is called an *interior algorithm* (or *midpoint algorithm* if $m(f)$ is a *midpoint* for every f).

Remark B.1 We motivate our terminology. For a given $A \subset F$ define

$$\operatorname{conv} S(A) = \{g \in G: \exists f_1,f_2 \in A, \operatorname{dist}_G(f_1,g) + \operatorname{dist}_G(f_2,g) = \operatorname{dist}(f_1,f_2)\}.$$

We shall say that $S(A)$ is *convex* iff $S(A) = \operatorname{conv} S(A)$. Note that for many problems $\operatorname{conv} S(A)$ is the convex hull of $S(A)$, and our definition of convexity coincides with the classical definition. For example, this is true for the linear normed space setting when the norm $\|\cdot\|$ is strictly convex. (The norm $\|\cdot\|$ is strictly convex iff $\|g_1 + g_2\| = \|g_1\| + \|g_2\|$ and $g_2 \neq 0$ implies that $g_1 = tg_2$ for some $t \in \mathbb{R}_+$.) Thus, $\operatorname{conv} S(A)$ is a generalization of the convex hull of $S(A)$.

It is easy to observe that every interior point m of $S(A)$ belongs to $\operatorname{conv} S(A)$, and if $S(A)$ is convex, then $m \in S(A)$. This justifies the name interior point.

For the normed linear space setting with a strictly convex norm, if m is a midpoint of $S(A)$, then $S(A)$ is symmetric with respect to m (that is, $g \in S(A)$ implies that $2m - g \in S(A)$). Thus, roughly speaking, m is in the "middle" of $S(A)$.

■

We illustrate interior and midpoint algorithms by the following example.

Example: Binary Search (continued) Since $F \subset G$, then $S(A) = A$. It is easy to observe that

$$\text{dist}_G(f,g) = \inf\{\delta: |f - g| \leq \delta m\} = \frac{|f - g|}{m}$$

and

$$\text{dist}(f_1,f_2) = \frac{|f_1 - f_2|}{m} .$$

Let $\rho = 0$ and $N(f) = [Q(f;T_1),\ldots, Q(f;T_n)]$. Let $a \in V(N,f)$. Setting $P(f) \equiv a$, we have

$$\text{dist}_G(\tilde{f},a) + \text{dist}_G(a,a) = \text{dist}(\tilde{f},a), \qquad \forall \tilde{f} \in V(N,f),$$

which proves that every element a from $V(N,f)$ is interior. Let ϕ^I be an interpolatory algorithm. Then $\phi^I(N(f)) \in V(N,f)$, which means that ϕ^I is an interior algorithm and

$$e(\phi^I,N,f) \leq 2r(N,f).$$

Does there exist a midpoint algorithm ϕ? Observe that ϕ exists iff $V(N,f)$ is symmetric, i.e.,

$$\tilde{f} \in V(N,f) \quad \text{implies} \quad 2c - \tilde{f} \in V(N,f)$$

for some $c = c(f)$ from F. Assume then that such a $c(f)$ exists. Then $\phi(N(f)) = c(f)$ is a midpoint algorithm. Note that

$$e(\phi,N,f) = r(N,f),$$

which yields that ϕ is also central. ■

We noticed that for the Binary Search Problem every interior algorithm is almost strongly optimal and every midpoint algorithm is central. We prove that this is true in general.

THEOREM B.1 Let ϕ be an interior algorithm. Then

$$e(\phi,N_\rho,f) \leq 2r(N_\rho,f), \qquad \forall f \in F. \tag{B.7}$$

Furthermore, if ϕ is a midpoint algorithm, then

$$e(\phi,N_\rho,f) = r(N_\rho,f), \qquad \forall f \in F, \tag{B.8}$$

i.e., ϕ is central. ■

PROOF Let $f \in F$ and $\phi(N_\rho(f)) = m$, where m is an interior point of $S(V(N_\rho,f))$. Then

$$\begin{aligned} e(\phi,N_\rho,f) &= \sup_{\tilde{f} \in V(N_\rho,f)} \inf\{\delta : m \in S(\tilde{f},\delta)\} \\ &= \sup_{\tilde{f} \in V(N_\rho,f)} \mathrm{dist}_G(\tilde{f},m) \\ &\le \sup_{\tilde{f} \in V(N_\rho,f)} \{\mathrm{dist}_G(\tilde{f},m) + \mathrm{dist}_G(P(\tilde{f}),m)\} \\ &= \sup_{\tilde{f} \in V(N_\rho,f)} \mathrm{dist}(\tilde{f},P(\tilde{f})). \end{aligned}$$

From (B.3) we get that $\mathrm{dist}(\tilde{f},P(\tilde{f})) \le 2r(N_\rho,f)$, which implies that $e(\phi,N_\rho,f) \le 2r(N_\rho,f)$. Thus, (B.7) is proven.

Let now ϕ be a midpoint algorithm. Then for every $\tilde{f}$ from $V(N_\rho,f)$ we have

$$\mathrm{dist}_G(\tilde{f},m) = \tfrac{1}{2}\mathrm{dist}(\tilde{f},P(\tilde{f})) \le r(N_\rho,f),$$

which implies that $e(\phi,N_\rho,f) \le r(N_\rho,f)$. Since $e(\phi,N_\rho,f) \ge r(N_\rho,f)$, (B.8) is proven. This completes the proof of Theorem B.1. ■

Theorem B.1 states that *interior algorithms* (if they exist) are *almost strongly optimal* and *midpoint algorithms* (if they exist) *are central.*

We now exhibit relations between interior and interpolatory algorithms. From Remark B.1 we know that if $S(V(N_\rho,f))$ is convex, then $\phi(N_\rho(f)) \in S(V(N_\rho,f))$ for every interior algorithm ϕ. Thus, ϕ is also an interpolatory algorithm, $\phi \in \Phi^I(N_\rho)$.

We prove that for regular S the opposite result holds. The problem S is *regular* iff for every $f_1,f_2 \in F$ we have

$$S(f_1,0) \subset S(f_2, \mathrm{dist}(f_1,f_2) + \delta), \qquad \forall \delta > 0. \tag{B.9}$$

THEOREM B.2 Let N_ρ be an approximate information operator. Then,

(i) If $S(V(N_\rho,f))$ is convex, $\forall f \in F$, then every interior algorithm is interpolatory.

(ii) If S is regular, then every interpolatory algorithm is interior. ■

PROOF

(i) This was established above.

(ii) Let $\phi \in \Phi^I(N_\rho)$. Then for every $f \in F$ there exists $f^* \in V(N_\rho,f)$ such that

$$\phi^I(N_\rho(f)) = g \in S(f^*,0).$$

We prove that g is an interior point of $S(V(N_\rho,f))$. Let

$$P(\tilde{f}) \equiv f^*, \qquad \forall \tilde{f} \in V(N_\rho,f).$$

Then

$$\text{dist}_G(\tilde{f},g) + \text{dist}_G(P(\tilde{f}),g) = \text{dist}_G(\tilde{f},g), \qquad \forall \tilde{f} \in V(N_\rho,f). \quad \text{(B.10)}$$

Due to (B.9),

$$\text{dist}_G(\tilde{f},g) = \inf\{\delta : g \in S(\tilde{f},\delta)\} \leq \text{dist}(\tilde{f},f^*) = \text{dist}(\tilde{f},P(\tilde{f})).$$

Thus

$$\text{dist}_G(\tilde{f},g) + \text{dist}_G(P(\tilde{f}),g) \leq \text{dist}(\tilde{f},P(\tilde{f})).$$

Since $\text{dist}_G(f_1,g) + \text{dist}_G(f_2,g) \geq \text{dist}(f_1,f_2)$, $\forall f_1,f_2 \in F$, $\forall g \in G$, we have

$$\text{dist}_G(\tilde{f},g) + \text{dist}_G(P(\tilde{f}),g) = \text{dist}(\tilde{f},P(\tilde{f})), \qquad \forall \tilde{f} \in V(N_\rho,f).$$

This means that $g = \phi^I(N_\rho(f))$ is an interior point of $S(V(N_\rho,f))$, which implies that ϕ^I is an interior algorithm. This completes the proof. ■

Theorem B.2 says that for regular problems every interpolatory algorithm is interior and therefore it is almost strongly optimal. Since interpolatory algorithms are conceptually simple, this is a desirable property.

We end this appendix by two examples which illustrate the concept of regularity.

Example: Normed Linear Space (continued) For this problem we have $S(f,\delta) = \{g \in G: \| S(f) - g \| < \delta\}$. Furthermore

$$\text{dist}_G(f,g) = \| S(f) - g \|$$

and

$$\text{dist}(f_1,f_2) = \| S(f_1) - S(f_2) \|.$$

For any $f_1, f_2 \in F$ and $\delta > 0$ we have

$$S(f_1,0) = \{S(f_1)\} \subset \{g \in G: \| S(f_2) - g \| < \| S(f_1) - S(f_2) \| + \delta\}$$

$$= S(f_2, \text{dist}(f_1,f_2) + \delta).$$

Thus, this problem is regular.

Note that this problem is also regular when G is equipped with a metric μ instead of a norm, i.e.,

$$S(f,\delta) = \{g \in G: \mu(S(f),g) \leq \delta\}. \qquad \blacksquare$$

Example: Nonlinear Equations Let F be a class of scalar functions f, $f:[0,1] \to \mathbb{R}$ that have a unique zero $\alpha, \alpha = \alpha(f)$. Our problem is to approximate $\alpha(f)$, that is, $G = \mathbb{R}$ and

$$S(f,\epsilon) = \{x \in G: | x - \alpha | \leq \epsilon, \text{ where } f(\alpha) = 0\}.$$

Note that $S(f,0)$ is a singleton, i.e.,

$$S(f,0) = \{\alpha(f)\}, \qquad \forall f \in F. \tag{B.11}$$

The problem S is regular. Indeed, for $f_1, f_2 \in F$ we have

$$\text{dist}(f_1,f_2) = | \alpha(f_1) - \alpha(f_2) |,$$

and therefore $\alpha(f_1) \in S(f_2, \text{dist}(f_1,f_2) + \delta)$, $\forall \delta > 0$, which yields regularity of S.

The assumption on uniqueness of zeros is crucial for regularity. To show this, assume that f from F can have many zeros. Taking $f_1(x) = x(x - 1)$ and $f_2(x) = x - 1/3$, we have

$$S(f_1,0) = \{0,1\}, \qquad S(f_2,0) = \{1/3\}.$$

Then dist $(f_1,f_2) = 1/3$. On the other hand,

$$S(f_2, \text{dist}(f_1,f_2) + \delta) = [-\delta, 2/3 + \delta],$$

which means that $S(f_1,0) \not\subset S(f_2, \text{dist}(f_1, f_2) + \delta)$ for $\delta \in (0,1/3)$. Thus S is not regular in this case. $\blacksquare$

APPENDIX C

DIAMETER OF INFORMATION

For the linear normed space setting it was shown that the diameter of information, which differs from the radius by at most a factor of two, is very useful and often much easier to compute (see GTOA). The diameter $d(N)$ of information N is defined as

$$d(N) = \sup_{f \in F} \text{diam } S(V(N,f)) \tag{C.1}$$

where

$$\text{diam } S(V(N,f)) = \sup_{f_1,f_2 \in V(N,f)} \| S(f_1) - S(f_2) \|$$

is the diameter of the set $S(V(N,f))$. Since rad $A \leq$ diam $A \leq 2$ rad A, for every subset A of G, then

$$r(N) \leq d(N) < 2\, r(N). \tag{C.2}$$

Furthermore, for every interpolatory algorithm $\phi^I \in \Phi(N)$, we have

$$e(\phi^I,N) \leq d(N). \tag{C.3}$$

Can the diameter of information be generalized so that it enjoys properties similar to those in the normed linear space setting, i.e., so that it is easier to compute than the radius and satisfies (C.2), (C.3)?

We provide an affirmative answer under a regularity assumption on S. At the end of the appendix we prove that the regularity assumption is essential. We use the notation of Appendix B. Let $A \subset F$ and $S(A) = S(A,0) = \bigcup_{f \in A} S(f,0)$. Then

$$\text{diam } S(A) = \sup_{f_1,f_2 \in A} \text{dist}(f_1,f_2)$$

is called the *diameter of the set* $S(A)$.

Let N_ρ be an approximate information operator. Then

$$d(N_\rho,f) = \text{diam } S(V(N_\rho,f)) \tag{C.4}$$

is the *local diameter of* N_ρ and

$$d(N_\rho) = \sup_{f \in F} d(N_\rho,f)$$

is the *global diameter of* N_ρ.

To establish relations between diameters and radii of information observe that

$$r(N_\rho,f) = \text{rad } S(V(N_\rho,f)), \tag{C.5}$$

where the radius of a set $S(A)$, $A \subset F$, is defined as

$$\text{rad } S(A) = \inf_{g \in G} \sup_{f \in A} \text{dist}_G(f,g) \qquad (= \inf\{\delta: \bigcap_{f \in A} S(f,\delta) \neq \emptyset\}).$$

We are ready to prove

LEMMA C.1 Let $A \subset F$. Then

$$\text{diam } S(A) \leq 2 \text{ rad } S(A). \tag{C.6}$$

If S is regular (see (B.9)), then

$$\text{rad } S(A) \leq \text{diam } S(A). \tag{C.7}$$

■

PROOF Suppose that $r = \text{rad } S(A)$ is finite. Let $\delta = r + \eta$ with $\eta > 0$. Take f_1, f_2 from A. Then $S(f_1,\delta) \cap S(f_2,\delta) \neq \emptyset$. This yields that

$$\text{dist}(f_1,f_2) = \inf\{\delta_1 + \delta_2 : S(f_1,\delta_1) \cap S(f_2,\delta_2) \neq \emptyset\} \leq 2\delta.$$

Thus, $\text{diam } S(A) \leq 2(r + \eta)$. Since η is arbitrary, (C.6) is proven.

To prove (C.7) assume that $d = \text{diam } S(A)$ is finite. Let $\delta = d + \eta$ with $\eta > 0$. Take f_1 from A. Then for every $f \in A$ we have dist $(f_1,f) < \delta$. Regularity of S yields $S(f_1,0) \subset S(f,d)$. Thus

$$\bigcap_{f \in A} S(f,\delta) \neq \emptyset,$$

which means that $\text{rad } S(A) \leq \delta = d + \eta$. Since η is arbitrary, $\text{rad } S(A) \leq d$, which completes the proof. ■

It is easy to observe that both inequalities (C.6) and (C.7) are sharp, i.e., there exist a problem S and sets A_1, A_2 such that diam $S(A_1) = 2$ rad $S(A_1)$ and rad $S(A_2) =$ diam $S(A_2)$. Indeed, let

$$F = G = \{1,2,3\} \qquad \text{and} \qquad S(i,\delta) = \{g \in G: \mid i - g \mid \leq \delta\}, \quad i \in F.$$

It is easy to see that S is regular. Take $A_1 = F$ and $A_2 = \{1,2\}$. Then diam $S(A_1)$ $= 2 = 2$ rad $S(A_1)$ and rad $S(A_2) = 1 =$ diam $S(A_2)$.

We also show that regularity of S is essential for (C.7). That is, there exist a nonregular problem S and $A \subset F$ such that rad $S(A) = +\infty$ and diam $S(A)$ $= 0$. Indeed, let $F = G = \{1,2,3\}$ and S be defined as follows:

$$S(1,\delta) = \{1,2\}, \qquad S(2,\delta) = \{2,3\}, \qquad S(3,\delta) = \{1,3\}.$$

Note that S is not regular since $S(1,0) = \{1,2\} \not\subset S(3,\delta) = \{1,3\}$, $\forall\delta \geq 0$. For $A = F$ we have

$$S(i,0) \cap S(j,0) \neq \emptyset, \quad \forall i,j \in F.$$

Thus, diam $S(A) = 0$. To prove that rad $S(A) = +\infty$, observe that

$$S(1,\delta) \cap S(2,\delta) \cap S(3,\delta) = \emptyset, \qquad \forall\ \delta \in \mathbb{R}_+.$$

Thus, rad $S(A) = +\infty$.

From Lemma C.1 we easily obtain

THEOREM C.1 Let N_ρ be an approximate information operator. Then

$$d(N_\rho,f) \leq 2r(N_\rho,f), \qquad \forall f \in F,$$
$$d(N_\rho) \leq 2\, r(N_\rho). \tag{C.8}$$

If S is regular, then

$$r(N_\rho,f) \leq d(N_\rho,f), \qquad \forall f \in F, \qquad \text{and} \qquad r(N_\rho) \leq d(N_\rho). \tag{C.9}$$

■

Note that (C.8) and (C.9) generalize (C.2). Theorem C.1 states that for regular problems the diameter of information is closely related to the radius.

For the normed linear space setting, which is a special case of a regular problem, we know that the local error of any interpolatory algorithm is no greater than $d(N_\rho,f)$. We prove that this property holds for any regular problem. We first establish the corresponding result for interior algorithms.

LEMMA C.2 Let ϕ be an interior algorithm. Then

$$e(\phi,N_\rho,f) \le d(N_\rho,f), \qquad \forall f \in F. \qquad \blacksquare$$

PROOF Let $f \in F$ and $m = \phi(N_\rho(f))$ be an interior point of $S(V(N_\rho,f))$. Then

$$e(\phi,N_\rho,f) = \sup_{f_1 \in V(N_\rho,f)} \operatorname{dist}_G(f_1,m) \le \sup_{f_1 \in V(N_\rho,f)} \operatorname{dist}(f_1,P(f_1)).$$

Since $P(f_1) \in V(N_\rho,f)$, then

$$e(\phi,N_\rho,f) \le \sup_{f_1,f_2 \in V(N_\rho,f)} \operatorname{dist}(f_1,f_2) = \operatorname{diam} S(V(N_\rho,f)) = d(N_\rho,f).$$

This completes the proof. $\blacksquare$

From Theorem B.2 we know that for regular problems every interpolatory algorithm is interior. Thus, due to Lemma C.2, we have

THEOREM C.2 Let N_ρ be an approximate information operator. If S is regular, then

$$e(\phi^I,N_\rho,f) \le d(N_\rho,f), \qquad \forall f \in F, \tag{C.10}$$

for every interpolatory algorithm $\phi^I \in \Phi^I(N_\rho)$. $\blacksquare$

We show that regularity of S is essential for (C.10). Above, we exhibited a problem for which diam $S(F) = 0$ and rad $S(F) = +\infty$. Take $N \equiv 0$. Then $S(V(N,f)) = G = F$, and $d(N,f) = 0$. On the other hand, for every algorithm ϕ we have $e(\phi,N,f) \ge r(N,f) = \text{rad } S(F) = +\infty$. Thus, for this problem we have

$$e(\phi,N,f) = +\infty \qquad \text{and} \qquad d(N,f) = 0.$$

APPENDIX D

NONDETERMINISTIC ALGORITHMS

In this appendix we introduce the concept of nondeterministic algorithm. We will obtain tight bounds on the power of nondeterminism. We first define nondeterministic algorithm in our setting and then compare and contrast our concept with that used in the theory of NP-completeness.

So far, we have considered algorithms whose values are single elements of the set G. That is, if we assume for simplicity that N is an exact information operator and ϕ is an algorithm that uses N, then $\phi(N(f))$ is an element of G. If

$$\phi(N(f)) \in S(\tilde{f},\epsilon), \qquad \forall \tilde{f} \in V(N,f), \tag{D.1}$$

then $\phi(N(f))$ is an ϵ-approximation.

In this appendix we generalize the concept of algorithm by assuming that $\phi(N(f))$ is a subset of G that consists of at most k elements:

$$\text{card}(\phi(N(f)) \leq k, \qquad \forall f \in F, \tag{D.2}$$

where k is a fixed integer. We then generalize (D.1) as follows. We say that the algorithm ϕ supplies an ϵ-approximation if for every $\tilde{f}$ from $V(N,f)$ there exists an element of $\phi(N(f))$ belonging to $S(\tilde{f},\epsilon)$. This means that

$$\phi(N(f)) \cap S(\tilde{f},\epsilon) \neq \emptyset, \qquad \forall \tilde{f} \in V(N,f). \tag{D.3}$$

For $k > 1$, we shall call such algorithms *nondeterministic*. Such an algorithm is said to be nondeterministic because (D.3) does not specify a unique element.

We now discuss the famous problem whether P = NP, and contrast it with our setting. We follow the treatment of Garey and Johnson [79]. The formal setting involves deterministic and nondeterministic Turing machines; we present an informal description.

A nondeterministic algorithm consists of two stages: "guessing" and "checking." Guessing is nondeterministic, while checking takes deterministic polynomial time. Together, the two stages take nondeterministic polynomial time. The class of problems that can be solved in nondeterministic polynomial time is denoted by NP, while the class of problems that can be solved in deterministic polynomial time is denoted by P. The central question is whether P = NP. It is widely believed that the hardest problems in NP (the NP-complete problems) have exponential deterministic time complexity and that therefore the two classes are very different.

We compare the above with our notion of nondeterminism. The fact that $\psi(N(f))$ can consist of k elements corresponds to guessing, while (D.3) corresponds to checking. The differences include the following:

1. Since we seek to understand the trade-off between nondeterminism and information, we do not restrict ourselves to a class of realizable algorithms (i.e., Turing machine algorithms).

2. In our general setting we do not wish to assign a cost to (D.3). Therefore we do not present complexity results. (However for particular applications our results could be used to obtain complexity bounds.)

3. We obtain very tight bounds on the trade-off between k-nondeterminism, on the one hand, and cardinality and radius of information, on the other (see Corollary D.2 and Lemma D.2). For example, Lemma D.2 establishes the existence of linear problems for which k-nondeterminism does not help, as well as the existence of other linear problems for which k-nondeterminism reduces the radius of information by k.

For an exact information operator N and $k \geq 1$, we define the k-nondeterministic local radius $r_k(N,f)$ of information as follows. We decompose the set $V(N,f) = \{\tilde{f} \in F\colon N(\tilde{f}) = N(f)\}$ as the union of k subsets of F:

$$V(N,f) = V_1 \cup V_2 \cup \cdots \cup V_k \tag{D.4}$$

in such a way that $\bigcap_{\tilde{f} \in V_i} S(\tilde{f},\delta)$ is nonempty for every i with a minimal δ. This minimal δ is the k-radius $r_k(N,f)$. More precisely,

$$\begin{aligned} r_k(N,f) = \inf\{\delta\colon \exists\, V_1,\ldots, V_k,\ V(N,f) = V_1 \cup \cdots \cup V_k, \\ \bigcap_{\tilde{f} \in V_i} S(\tilde{f},\delta) \neq \emptyset,\ i = 1,\ldots,k\} \end{aligned} \tag{D.5}$$

is called the *k-nondeterministic local radius of information* (briefly, the k-local radius). The *k-nondeterministic (global) radius of information* (briefly, the k-global radius) is defined as

$$r_k(N) = \sup_{f \in F} r_k(N,f). \tag{D.6}$$

Obviously, $r_k(N) \le r_{k-1}(N) \le r_1(N) = r(N)$, $\forall k > 1$, $\forall N$.

We show that the k-local and global radii are the sharp lower bounds on the local and global errors of k-nondeterministic algorithms. The *local error of a k-nondeterministic algorithm* ϕ is defined as

$$e(\phi,N,f) = \inf\{\delta\colon \phi(N(f)) \cap S(\tilde{f},\delta) \ne \emptyset,\ \forall \tilde{f} \in V(N,f)\} \tag{D.7}$$

and the *global error* as

$$e(\phi,N) = \sup_{f \in F} e(\phi,N,f). \tag{D.8}$$

Let $\Phi_k(N)$ be the class of all k-nondeterministic algorithms.

THEOREM D.1

$$\inf_{\phi \in \Phi_k(N)} e(\phi,N,f) = r_k(N,f), \qquad \forall f \in F, \tag{D.9}$$

$$\inf_{\phi \in \Phi_k(N)} e(\phi,N) = r_k(N). \tag{D.10}$$

■

PROOF Take ϕ from $\Phi_k(N)$ and assume without loss of generality that $e(\phi,N,f)$ is finite. Let $\delta > e(\phi,N,f)$. Then $\phi(N(f)) \cap S(\tilde{f},\delta) \ne \emptyset$, $\forall \tilde{f} \in V(N,f)$. Let $A = \{x_1,x_2,\ldots,x_p\}$ be the set of all elements from $\phi(N(f))$ such that x_i belongs to $S(\tilde{f},\delta)$ for some $\tilde{f} \in V(N,f)$, $i = 1,2,\ldots,p$. Since ϕ is k-nondeterministic, $p \le k$. Define

$$V_i = \{\tilde{f} \in V(N,f) : x_i \in S(\tilde{f},\delta)\}, \qquad i = 1,2,.., p,$$

$$V_{p+i} = V_p, \qquad i = 1,2,\ldots, k-p.$$

Then V_i is nonempty, $V(N,f) = V_1 \cup \cdots \cup V_k$ and $\bigcap_{\tilde{f} \in V_i} S(\tilde{f},\delta) \ne \emptyset$. From (D.5) we get $r_k(N,f) \le \delta$. Letting δ tend to $e(\phi,N,f)$, we have $r_k(N,f) \le e(\phi,N,f)$. Since ϕ is arbitrary,

$$\inf\{e(\phi,N,f) : \phi \in \Phi_k(N)\} \ge r_k(N,f).$$

To prove the reverse inequality, assume without loss of generality that $r_k(N,f)$ is finite. Take $\delta > r_k(N,f)$. Then (D.5) yields that there exist $V_1, V_2,\ldots, V_k$ such that $V(N,f) = V_1 \cup \cdots \cup V_k$ and $\bigcap_{\tilde{f} \in V_i} S(\tilde{f},\delta)$ is nonempty for $i = 1,2,\ldots,k$. Let

$$x_i \in \bigcap_{\tilde{f} \in V_i} S(\tilde{f},\delta).$$

Define $\phi_\delta(N(f)) = \{x_1, x_2, \ldots, x_k\}$. Then ϕ_δ is k-nondeterministic and $\phi_\delta(N(f)) \cap S(\tilde{f},\delta) \neq \emptyset$, $\forall \tilde{f} \in V(N,f)$. Thus $e(\phi_\delta,N,f) \leq \delta$. Letting δ tend to $r_k(N,f)$, we get

$$\inf e(\phi_\delta,N,f) \leq r_k(N,f).$$

This completes the proof of (D.9). Condition (D.10) easily follows from (D.9). ■

In Chapter 5 we defined the ϵ-entropy of X, $X \subset F$, as $H(\epsilon,X) = \log_2 k$, where k is the minimal number of balls with radii not greater than ϵ that cover the set X (see (4.16) of Chapter 5). By the ball we mean

$$J(g,\delta) = \{f \in F\colon \operatorname{dist}_G(f,g) \leq \delta\} = \{f \in F\colon g \in S(f,\delta + \eta),\ \forall \eta > 0\}.$$

By $H^{-1}(k,X)$ we mean the infimum of ϵ such that $\log_2 k \geq H(\epsilon,X)$. We now study how the k-nondeterministic local and global radii are relaed to ϵ-entropy.

LEMMA D.1

$$r_k(N,f) = H^{-1}(k,V(N,f)), \qquad \forall f \in F, \tag{D.11}$$

$$r_k(N) = \sup_{f \in F} H^{-1}(k,V(N,f)). \tag{D.12}$$

■

PROOF Let $a = H^{-1}(k,V(N,f))$. Take $\phi \in \Phi_k(N)$. Then

$$\phi(N(f)) = \{x_1,\ldots,x_k\} \qquad \text{and} \qquad V(N,f) \subset \bigcup_{i=1}^{k} J(x_i,e(\phi,N)).$$

Thus $a \leq e(\phi,N)$. Taking the infimum with respect to ϕ, we get $a \leq r_k(N,f)$.

To prove the reverse inequality for $a < +\infty$, take $\delta > a$. From the definition of H^{-1}, there exist $g_1,\ldots,g_k$ such that

$$V(N,f) \subset \bigcup_{i=1}^{k} J(g_i,\delta).$$

Then the algorithm $\phi(N(f)) = \{g_1,\ldots,g_k\}$ belongs to $\Phi_k(N)$ and $\phi(N(f)) \cap S(\tilde{f},\delta + \eta) \neq \emptyset$, $\forall \tilde{f} \in V(N,f)$ and $\forall \eta > 0$. Thus $e(\phi,N,f) \leq \delta$. Letting δ tend to a, we get $r_k(N,f) \leq a$. This completes the proof of (D.11). Condition (D.12) easily follows from (D.11). ■

We now turn to the problem of optimal information for the nondeterministic case. As in Chapter 4, let $\Psi(H_1)$ be a class of exact primitive information operators $U\colon F \to H_1$, let $\Psi^{\text{non}}(n)$ be the class of all nonadaptive information with cardinality at most n, and let $\Psi^a(n)$ be the corresponding class of adaptive information. We define the *nth minimal k-nondeterministic radius of information* $r_k^{\text{non}}(n)$ for nonadaptive information and $r_k^a(n)$ for adaptive information as

$$r_k^{\text{non}}(n) = \inf_{N \in \Psi^{\text{non}}(n)} r_k(N), \qquad r_k^a(n) = \inf_{N \in \Psi^a(n)} r_k(N). \tag{D.13}$$

Similarly, the *k-nondeterministic ϵ-cardinality numbers* for nonadaptive and adaptive information are defined as

$$m_k^{\text{non}}(\epsilon) = \min\{n\colon r_k^{\text{non}}(n) \le \epsilon\},\ m_k^a(\epsilon) = \min\{n\colon r_k^a(n) \le \epsilon\}. \tag{D.14}$$

We exhibit some relations between the minimal radii for the nondeterministic case ($k > 1$) and for the deterministic case ($k = 1$).

We begin with discrete information operators. That is, as in section (iii) of Chapter 5, we assume that H_1 is a set consisting of p elements, $p > 1$, and $\Psi(H_1) = \bar{\Psi}(H_1)$, i.e., the class of primitive information operators consists of *all* mappings (generalized questions) $U\colon F \to H_1$. We now generalize Theorem 4.1 of Chapter 5.

THEOREM D.2

$$r_k^{\text{non}}(n) = r_k^a(n) = H^{-1}(kp^n,F), \tag{D.15}$$

$$m_k^{\text{non}}(\epsilon) = m_k^a(\epsilon) = \max\left(\left\lceil \frac{H(\epsilon,F) - \log k}{\log p} \right\rceil, 0\right). \tag{D.16}$$

■

PROOF From Theorem (3.1) of Chapter 4 it follows that $\Psi(H_1) = \bar{\Psi}(H_1)$ implies equality between classes of nonadaptive and adaptive information operators. This yields that $r_k^{\text{non}}(n) = r_k^a(n)$ and $m_k^{\text{non}}(\epsilon) = m_k^a(\epsilon)$.

Let $N = [U_1,\ldots,U_n]$ with $r_k(N) < +\infty$. We prove that $a = H^{-1}(kp^n,F) \le r_k(N)$. Let $\eta > 0$. Take an algorithm $\phi \in \Phi_k(N)$ such that $e(\phi,N) < r_k(N) + \eta$. Since $N(F)$ has at most p^n distinct elements and $\phi(N(f))$ has at most k distinct elements, then

$$\bigcup_{f \in F} \{\phi(N(f))\} = \{g_1,\ldots,g_j\} \quad \text{with} \quad j \le kp^n.$$

Note that $\phi(N(f)) \cap S(f,r_k(N) + \eta) \ne \emptyset$, $\forall f \in F$, which implies that

$$F \subset \bigcup_{i=1}^{j} J(g_i,r_k(N) + \eta).$$

Thus $a \le r_k(N) + \eta$. Since η can be arbitrarily close to zero and N is any information operator from $\Psi^a(n)$, we get $a \le \inf\{r_k(N) : N \in \Psi^a(n)\}$.

To prove the reverse inequality, assume that a is finite and take elements $g_1,\ldots,g_j$, $j = kp^n$, such that $F \subset \bigcup_{i=1}^{j} J(g_i,a + \delta)$ for some $\delta > 0$. Let

$$A_i = \{g_{(i-1)k+1},\ldots,g_{(i-1)k+k}\}, \text{ i } = 1,\ldots,p^n,$$

and let P be a one to one mapping of $\{A_1,\ldots,A_{p^n}\}$ into H_1^n. Define the information operator

$$N(f) = PA_i, \quad \text{where} \quad f \in \bigcup_{g \in A_i} J(g,a + \delta). \tag{D.17}$$

Note that $N \in \Psi^{\text{non}}(n)$. Define the algorithm

$$\phi(N(f)) = P^{-1}N(f) = A_i. \tag{D.18}$$

Then $\phi \in \Phi_k(N)$ and $\phi(N(f)) \cap S(f,a + \eta) \neq \emptyset$, $\forall f \in F$, $\forall \eta > \delta$. Thus $r_k(N) \leq e(\phi,N) \leq a + \delta$. Since δ is arbitrary, $\inf\{r_k(N):N \in \Psi^a(n)\} \leq a$, and (D.15) is proven.

To prove (D.16), observe that n is the minimal nonnegative solution of $\log kp^n \geq H(\epsilon,F)$. Thus, $n \geq [H(\epsilon,F) - \log k]/\log p$, which yields (D.16) and completes the proof. ■

From the proof of Theorem D.2 we conclude

COROLLARY D.1 The information N defined by (D.17) is nth optimal and the algorithm ϕ defined by (D.18) is an optimal error algorithm, i.e.,

$$r_k(N) = r_k^a(n), \qquad e(\phi,N) = r_k(N).$$

■

Theorem D.2 enables us to compare k-nondeterminism with the deterministic case $k = 1$. From (D.15) and (D.16) we have

$$\begin{gathered} r^a(n + c - 1) \leq r_k^a(n) \leq r^a(n + c), \\ \max(m^a(\epsilon) - c,0) \leq m_k^a(\epsilon) \leq \max(m^a(\epsilon) - c + 1,0), \end{gathered} \tag{D.19}$$

where $c = \lceil \log k/\log p \rceil$. This means that by using the best information operator with cardinality $(n + c - 1)$ or $(n + c)$ we can get the same error as by using the best information of cardinality n for the k-nondeterministic case. Thus we have

COROLLARY D.2 The k-nondeterminism saves at most $\lceil \log k/\log p \rceil$ evaluations of primitive information. ■

We now discuss information operators defined as for the linear normed space setting. Recall that $S:F_1 \to F_2$, $N:F_1 \to F_3$, $T:F_1 \to F_4$ are linear operators, where F_1 and F_3 are linear spaces, F_2 and F_4 are linear normed spaces, and $F = \{f \in F_1: \| Tf \| \le 1\}$.

Let $\phi(N(f)) = \{g_1(N(f)),\ldots,g_k(N(f))\}$, $\forall f \in F$. Then (D.3) easily yields

$$e(\phi,N) = \sup_{f \in F} \min_{1 \le i \le k} \| Sf - g_i(N(f)) \|. \tag{D.20}$$

LEMMA D.2

$$\frac{1}{2k}\, r(N) \le r_k(N) \le r(N). \tag{D.21}$$

■

PROOF We need to show only that $r_k(N) \ge r(N)/(2k)$. Recall that $r(N) = c \sup \{\| Sh \|: h \in \ker N, \| Th \| \le 1\}$ for some $c \in [1,2]$. Take an algorithm $\phi \in \Phi_k(N)$ and $h \in \ker N \cap$ F. Then $\phi(N(h)) = \phi(0) = \{g_1,g_2,\ldots,g_k\}$. Since $ch \in \ker N \cap F$ for $c \in [-1,1]$, there exists $g(c) \in \phi(0)$ such that $\| g(c) - cSh \| \le e(\phi)$. Let $c = c_i = -1 + 2i/k$, $i = 0,1,\ldots,k$. Since there are $(k + 1)$ different c_i and at most k different $g(c)$, there exists an element $g \in \phi(0)$ such that

$$\| g - c_iSh \| \le e(\phi), \qquad \| g - c_jSh \| \le e(\phi)$$

for some i and j, $i \ne j$. Then

$$\begin{aligned} 2e(\phi) &\ge \| g - c_iSh \| + \| g - c_jSh \| \ge \| c_iSh - c_jSh \| \\ &\ge \| Sh \| \min_{i \ne j} | c_i - c_j | = 2 \| Sh \| /k. \end{aligned}$$

Taking the supremum with respect to $h \in \ker N \cap F$, we get $e(\phi) \ge r(N)/(2k)$. Since ϕ is an arbitrary k-nondeterministic algorithm, $r_k(N) \ge r(N)/(2k)$, which completes the proof. ■

We now show that the left-hand side of (D.21) is sharp to within a factor of two and that the right-hand side is sharp. To show that the left-hand side is almost sharp, let S be a real linear functional. Then $S(V(N,f))$ is an interval $[a,b]$, and taking

$$\phi(N(f)) = \{a + h(2i + 1): i = 0,1,\ldots,k - 1,\ h = (b - a)/(2k)\}$$

we have $e(\phi,N,f) = (b - a)/(2k)$. Since $r(N,f) = (b - a)/2$, we conclude from (D.21) that

$$\tfrac{1}{2k} r(N) \le r_k(N) \le \tfrac{1}{k}\, r(N).$$

To prove that for some problems $r_k(N) = r(N)$, we proceed as follows. By a $(k + 1)$-dimensional ball in F_2 we understand the set

$$J(a,c) = \{g \in A: \| g - a \| \leq c\}, \qquad a \in F_2, \qquad c \in \mathbb{R}_+,$$

where $A = \text{lin}(z_1, z_2, \ldots, z_{k+1})$ for linearly independent elements z_i of F_2.

LEMMA D.3 Let $J(a,c)$ be a $(k + 1)$-dimensional ball in F_2 such that

$$J(a,c) \subset S(\ker N \cap F). \tag{D.22}$$

Then

$$r_k(N) \geq c. \tag{D.23}$$

■

PROOF Let ϕ be an arbitrary algorithm from $\Phi_k(N)$. Let $\phi(0) = \{g_1, g_2, \ldots, g_k\}$. We recall (see e.g. Kornejčuk [76] or Lorentz [66]) that the Kolmogorov k-width of a $(k + 1)$-dimensional ball of radius c is equal to c, i.e.,

$$d_k(J(a,c), F_2) \overset{df}{=} \inf_{A} \sup_{g \in J(a,c)} \inf_{x \in A} \| g - x \| = c,$$

where A is a linear subspace of F_2 of dimension at most k. Thus, in particular, there exists an element g of $J(a,c)$ such that $\| g - x \| \geq c$, $\forall x \in \text{lin}(g_1, g_2, \ldots, g_k)$. This implies that $\| g - g_i \| \geq c$, $i = 1,2,\ldots,k$. Note that (D.22) yields $g = Sf_0$ for $f_0 \in \ker N \cap F$. From (D.20) we get

$$e(\phi,N) \geq \min_{1 \leq i \leq k} \| Sf_0 - g_i \| \geq c.$$

Since ϕ is arbitrary, (D.10) yields $r_k(N) \geq c$. Hence (D.23) is proven.

■

As a simple example, observe that if $Sf = Tf = f$, $\forall f$, is the identity operator, and if $F_1 = F_2 = F_4$, then $S(\ker N \cap F)$ is a p-dimensional ball with center 0 and radius 1, where $p = \dim \ker N$. Note that $p = +\infty$ if N is a finite dimensional linear operator and $\dim F_1 = +\infty$. Observe also that $\phi(N(f)) \equiv 0$ is an optimal error algorithm if $p \geq 1$ and $e(\phi,N) = r(N) = 1$. Thus if $k + 1 \leq p$, then Lemma D.3 yields

$$r_k(N) = r(N) = 1.$$

As a final result in this appendix we show that sometimes a k-nondeterministic algorithm for the linear normed space setting may be replaced by a deterministic algorithm, with no higher error, that uses extra k evaluations of linear functionals.

Let F_2 be a Hilbert space. Then from (D.20) we have

$$e(\phi,N) \ge \min_i \| Sf - g_i(N(f)) \| \ge \min_{c_i} \| Sf - \sum_{i=1}^{k} c_i z_i \|$$

$$= \| Sf - \sum_{i=1}^{k} (Sf,z_i)z_i \|,$$

where $f \in F$ and $\text{lin}(g_1(N(f)),\ldots,g_k(N(f)) \subset \text{lin}(z_1,\ldots,z_k)$ and $z_1,\ldots,z_k$ are orthonormal. Let

$$N_1(f) = [N(f), (Sf,z_1),\ldots, (Sf,z_k)].$$

Since $z_i = z_i(N(f))$, N_1 is a linear adaptive information operator and

$$\text{card}(N_1) \le \text{card}(N) + k.$$

The algorithm

$$\phi_1(N_1(f)) = \sum_{i=1}^{k} (Sf,z_i)\, z_i$$

is deterministic and

$$e(\phi_1,N_1) \le e(\phi,N). \tag{D.24}$$

Since this holds for an arbitrary k-nondeterministic algorithm, we have proven

COROLLARY D.3 Let F_2 be a Hilbert space. Then k-nondeterminism can save at most k evaluations of linear functionals. ■

We now compare the nth minimal radii of nonadaptive and adaptive linear information operators of cardinality at most n.

If F_2 is a Hilbert space, then Corollary D.3 and Theorem 3.3 of Chapter 4 yield:

$$\frac{r^{\text{non}}(n+k)}{2} \le r^a(n+k) \le r_k^a(n) \le r_k^{\text{non}}(n)$$

$$\le r^{\text{non}}(n) \le 2r^a(n) \le 2r_k^a(n-k).$$

In particular, we have

$$a_1 \lim_n r^{\text{non}}(\text{n}) = a_2 \lim_n r_k^{\text{non}}(\text{n}) = \lim_n r^a(n) = \lim_n r_k^a(n),$$

with $a_2 \in [1/2,1]$ and $a_1/a_2 \in [1,2]$. These two relations can be summarized in the following corollary.

COROLLARY D.4 Let F_2 be a Hilbert space. Disregarding constants a_1 and a_2 we have:

(i) Adaption does not help for either the deterministic or nondeterministic case;

(ii) The limiting value of the minimal radii does not depend on nondeterminism;

(iii) If the problem is divergent from the deterministic case (i.e., $\lim_n r^a(n) > 0$), then it is divergent for the nondeterministic case as well. ∎

APPENDIX E

GENERALIZED NORMED LINEAR SPACE SETTING

In GTOA we studied the "linear problem" as a specific example of the normed linear space setting. The linear problem is defined as follows. Let $S: F_1 \rightarrow F_2$, $N: F_1 \rightarrow F_3$, $T: F_1 \rightarrow F_4$ be *linear operators,* where F_1 and F_3 are linear spaces, F_2 and F_4 are normed linear spaces. Let

$$F = \{f \in F_1 : \| Tf \| \leq 1\}, \tag{E.1}$$

$$S(f,\epsilon) = \begin{cases} \{g \in F_2 : \| g - S(f) \| < \epsilon\} & \text{if} \quad \epsilon > 0, \\ \{S(f)\} & \text{if} \quad \epsilon = 0. \end{cases} \tag{E.2}$$

To stress the dependence on the operators S and T, we shall write $r(N,S,T)$ instead of $r(N)$ as the radius of information. From Theorem C.1 we know that

$$r(N,S,T) = c\, d(N,S,T), \qquad c \in [1/2,1], \tag{E.3}$$

where the diameter $d(N,S,T)$ of information is given by

$$d(N,S,T) = \inf\{\delta_1 + \delta_2 : S(f_1,\delta_1) \cap S(f_2,\delta_2) \neq \emptyset, \quad \forall f_1,f_2 \in F,\ N(f_1) = N(f_2)\}. \tag{E.4}$$

It is easy to rewrite (E.4) as

$$d(N,S,T) = \sup_{f \in F} \sup_{f_1,f_2 \in V(N,f)} \| S(f_1) - S(f_2) \| = 2 \sup_{\substack{h \in \ker N \\ \| Th \| \leq 1}} \| Sh \|. \tag{E.5}$$

For some problems of practical importance it is difficult to verify whether $\| Tf \| \leq 1$. Furthermore, one may be interested in a different error criterion than (E.2). For instance, one may want to find an element g such that the relative error is small, i.e., $\| g - S(f) \| < \epsilon(\| S(f) \| + \eta)$ for some small nonnegative number η, or one may consider all f from F, i.e., the value of $\| Tf \|$ can be arbitrary, and redefine the error criterion as $\| g - S(f) \| / \max(\| Tf \|, 1) < \epsilon$.

This appendix deals with a generalization of (E.1), (E.2). Let

$$\rho : \mathbb{R}_+ \to \mathbb{R}_+ \tag{E.6}$$

be a given function. Let $F = F_1$, and consider the solution operator S_1 defined as

$$S_1(f,\epsilon) = \begin{cases} \{g \in F_2 : \| g - S(f) \| \rho(\| Tf \|) < \epsilon\} & \text{if } \epsilon > 0, \\ \{S(f)\} & \text{if } \epsilon = 0. \end{cases} \tag{E.7}$$

As with problem S, we shall write $r(N,S_1,T)$ and $d(N,S_1,T)$ to denote the radius and diameter of information.

We shall establish a rather surprising relation between the radii and diameters of the problems S and S_1. We first find the diameter of information for the problem S_1.

LEMMA E.1

$$d(N,S_1,T) = \sup_{N(f_1 - f_2) = 0} \| S(f_1 - f_2) \| \min(\rho(\| Tf_1 \|), \rho(\| Tf_2 \|)). \tag{E.8}$$

■

PROOF Recall that

$$d(N,S_1,T) = \inf\{\delta_1 + \delta_2 : S_1(f_1,\delta_1) \cap S_1(f_2,\delta_2) \neq \emptyset,$$

$$\forall f_1, f_2 \in F_1,\ N(f_1 - f_2) = 0\}.$$

For positive δ_1 and δ_2 we show that

$$S_1(f_1,\delta_1) \cap S_1(f_2,\delta_2) \neq \emptyset \text{ iff}$$

$$\| S(f_1 - f_2) \| < \frac{\delta_1}{\rho(\| Tf_1 \|)} + \frac{\delta_2}{\rho(\| Tf_2 \|)}, \tag{E.9}$$

with the convention $\delta_i/0 = +\infty$. Indeed, if $S_1(f_1,\delta_1) \cap S_1(f_2,\delta_2) \neq \emptyset$, then there exists an element g from F_2 such that

$$\| S(f_1) - g \| \rho(\| Tf_1 \|) < \delta_1 \quad \text{and} \quad \| S(f_2) - g \| \rho(\| Tf_2 \|) < \delta_2.$$

Hence

$$\| S(f_1 - f_2) \| \leq \| S(f_1) - g \| + \| S(f_2) - g \| < \frac{\delta_1}{\rho(\| Tf_1 \|)} + \frac{\delta_2}{\rho(\| Tf_2 \|)}.$$

Assume now that the right-hand side of (E.9) holds. If $\rho(\| Tf_2 \|) = 0$, set $g = Sf_1$. If $\rho(\| Tf_1 \|) = 0$, set $g = Sf_2$. Then $g \in S_1(f_1,\delta_1) \cap S_1(f_2,\delta_2)$ and therefore $S_1(f_1,\delta_1) \cap S_1(f_2,\delta_2)$ is nonempty. If $Sf_1 = Sf_2$, then $g = Sf_1 \in S_1(f_1,\delta_1) \cap S_1(f_2,\delta_2)$. Thus we can assume that $\rho(\| Tf_i \|) \neq 0$ for $i = 1,2$ and $S(f_1 - f_2) \neq 0$. Then set

$$g = tS(f_1) + (1 - t)\, S(f_2)$$

with $t = \delta_2(1 - x)/(\rho(\| Tf_2 \|)\, \| S(f_1 - f_2) \|)$, where x is defined by the equation

$$\| S(f_1 - f_2) \| = \frac{\delta_1(1 - x)}{\rho(\| Tf_1 \|)} + \frac{\delta_2(1 - x)}{\rho(\| Tf_2 \|)} .$$

Note that $x \in (0,1)$ and $t < 1$. We have

$$\| S(f_2) - g \| = \quad t\| S(f_1 - f_2) \| = \frac{\delta_2(1 - x)}{\rho(\| Tf_2 \|)} < \frac{\delta_2}{\rho(\| Tf_2 \|)} ,$$

$$\| S(f_1) - g \| = (1 - t)\| S(f_1 - f_2) \| = \frac{\delta_1(1 - x)}{\rho(\| Tf_1 \|)} < \frac{\delta_1}{\rho(\| Tf_1 \|)} .$$

Thus $g \in S(f_1,\delta_1) \cap S(f_2,\delta_2)$, which completes the proof of (E.9). Therefore

$$d(N,S_1,T) = \inf\{\delta_1 + \delta_2 : \| S(f_1 - f_2) \| \le \frac{\delta_1}{\rho(\| Tf_1 \|)} + \frac{\delta_2}{\rho(\| Tf_2 \|)} ,$$

$$\forall f_1,f_2 \in F_1,\ N(f_1 - f_2) = 0\}.$$

It is easy to check that

$$\inf\{\delta_1 + \delta_2 : a \le \delta_1/b + \delta_2/c\} = a \min(b,c)$$

for nonnegative a, b, and c. Setting $a = \| S(f_1 - f_2) \|$, $b = \rho(\| Tf_1 \|)$, and $c = \rho(\| Tf_2 \|)$, we get (E.8). ∎

We are ready to prove

THEOREM E.1 If $\sup\{\rho(x) : x \ge 0\} < +\infty$, then

$$r(N,S_1,T) = c\, d(N,S_1,T), \qquad c \in [1/2,1]. \tag{E.10}$$

If $\ker T \cap \ker N \subset \ker S$, then

$$d(N,S_1,T) = \sup_{x>0} x\,\rho(x)\,d(N,S,T) \tag{E.11}$$

with the convention $0 \cdot \infty = 0$. ∎

PROOF Since $r_1 = r(N,S_1,T) \geq d_1/2 = d(N,S_1,T)/2$ (due to Theorem C.1), we need to show only that $r_1 \leq d_1$. Assume then that $d_1 < +\infty$. Let $f \in F_1$ and let $a = \sup\{\rho(\| T\tilde{f} \|): \tilde{f} \in V(N,f)\}$. Since ρ is uniformly bounded, a is finite. For $\eta \in (0,1)$, take $f_1 \in V(N,f)$ such that $\rho(\| Tf_1 \|) \geq a(1 - \eta)$. Observe that

$$\rho(\| T\tilde{f} \|) \leq \frac{\rho(\| Tf_1 \|)}{1 - \eta}$$

and

$$\| S(f_1 - \tilde{f}) \|\, \rho(\| T\tilde{f} \|) = \| S(f_1 - \tilde{f}) \| \min\left(\rho(\| T\tilde{f} \|), \frac{\rho(\| Tf_1 \|)}{1 - \eta}\right),$$

$$\forall \tilde{f} \in V(N,f).$$

Since

$$\min\left(\rho(\| T\tilde{f} \|), \frac{\rho(\| Tf_1 \|)}{1 - \eta}\right) \leq \frac{\min(\rho(\| T\tilde{f} \|), \rho(\| Tf_1 \|))}{1 - \eta},$$

Lemma E.1 yields

$$\| S(f_1) - S(\tilde{f}) \|\, \rho(\| T\tilde{f} \|) \leq \frac{d_1}{1 - \eta}, \qquad \forall \tilde{f} \in V(N,f).$$

This means that $S(f_1)$ belongs to

$$\bigcap_{\tilde{f} \in V(N,f)} S_1\left(\tilde{f}, \frac{d_1}{1 - \eta}\right).$$

Since this holds for any f, we conclude that $r_1 \leq d_1/(1 - \eta)$. Letting η tend to zero, we get $r_1 \leq d_1$. This completes the proof of (E.10).

We now prove (E.11). Take any f_1 and f_2 such that $N(f_1 - f_2) = 0$. Let $x = \max(\| Tf_1 \|, \| Tf_2 \|)$. Then $h = f_1 - f_2 \in \ker N$ and $\| Th \| \leq 2x$. Let

$$\delta = \| S(f_1 - f_2) \| \min(\rho(\| Tf_1 \|), \rho(\| Tf_2 \|)). \tag{E.12}$$

If $x = 0$, then $h \in \ker T$, and the assumption $\ker T \cap \ker N \subset \ker S$ implies that $\delta = \| Sh \|\, \rho(0) = 0$. If $x \neq 0$, then

$$\delta \le 2\| S(\tfrac{1}{2x}h) \| \, x\, \rho(x) \le \Big(2 \sup_{\substack{h \in \ker N,\\ \|Th\| \le 1}} \| Sh \| \Big) \, x\, \rho(x)$$

$$\le \sup_{x > 0} x\, \rho(x)\, d(N,S,T)$$

due to (E.5). Since this holds for any f_1 and f_2, $N(f_1 - f_2) = 0$, (E.8) yields

$$d(N,S_1,T) \le \sup_{x > 0} x\, \rho(x)\, d(N,S,T).$$

To prove the reverse inequality, assume without loss of generality that $d = d(N,S,T) > 0$. Then $\ker T \cap \ker N \subset \ker S$ and (E.5) imply that $T \not\equiv 0$. Let $\eta > 0$. Take $h \in \ker N$ and $\| Th \| = 1$ such that $2\| Sh \| > d - \eta$ whenever d is finite or $2\| Sh \| > \eta$ otherwise. Define $f_1 = xh$, $f_2 = -xh$ for $x \in \mathbb{R}_+$. Then $N(f_1 - f_2) = 0$ and δ defined by (E.12) is equal to $\delta = 2\| Sh \| x\, \rho(x)$. If $d < +\infty$, then $d_1 = d(N,S_1,T) \ge \delta > (d - \eta)\, x\, \rho(x)$. If $d = +\infty$, then $d_1 \ge \eta\, x\, \rho(x)$. Since η and x are arbitrary, we get $d_1 \ge d\, x\, \rho(x)$ and $d_1 \ge d \sup_{x > 0} x\, \rho(x)$. This completes the proof.

■

Remark E.1 We comment on the assumptions of Theorem E.1. The first assumption $a = \sup\{\rho(x) : x \ge 0\} < +\infty$ was needed to prove that $r_1 \le d_1$. If $a = +\infty$, then it is possible to show that $r_1 \le d_1$ under some assumptions on the norm in F_4. We leave this case to the reader.

The second assumption $\ker T \cap \ker N \subset \ker S$ is a necessary condition for $d(N,S,T) < +\infty$. It is easy to show that $\ker T \cap \ker N \not\subset \ker S$ implies $d(N,S,T) = +\infty$ and

$$d(N,S_1,T) = \begin{cases} +\infty & \text{if } \rho(0) \ne 0 \text{ or } (\rho \not\equiv 0 \text{ and } T \not\equiv 0), \\ 0 & \text{otherwise.} \end{cases}$$

■

We illustrate Theorem E.1 by a number of specific functions ρ.

Example E.1 Let

$$\rho(x) = \begin{cases} 1 & \text{if } x \le 1, \\ 0 & \text{otherwise.} \end{cases}$$

Then $S_1(f,\epsilon) = S(f,\epsilon)$, $\forall f\colon \| Tf \| \le 1$. Thus the generalized linear problem coincides in this case with the linear problem (E.2). Note that (E.11) and Remark E.1 yield that $d(N,S_1,T) = d(N,S,T)$. ■

Example E.2 Let

$$\rho(x) = \frac{1}{x + \eta}, \qquad \eta > 0.$$

Then an ϵ-approximation g is an element such that

$$\| g - S(f) \| < \epsilon (\| Tf \| + \eta), \qquad \epsilon > 0.$$

Observe that for $T = S$ this is the relative error. In this case we also have $d(N,S_1,T) = d(N,S,T)$. ■

Example E.3 Let

$$\rho(x) = e^{(1 - x^2)/2}.$$

Then an ϵ-approximation g is an element such that

$$\| g - S(f) \| \, e^{-\| Tf \|^2/2} < \frac{\epsilon}{\sqrt{e}}, \qquad \epsilon > 0.$$

This means that g has to be close to $S(f)$ only if $\| Tf \|$ is relatively small. This example will be basic for the study of an average case model where the "weight" function $e^{-\| Tf \|^2/2}$ will be used to define a measure in the space F_1. Note that $\sup x \, \rho(x) = 1$, which yields $d(N,S_1,T) = d(N,S,T)$. ■

We now discuss relations between the errors of algorithms for the problems S and S_1. Let N be a linear information operator. Let $\phi : N(F) \to F_2$ be an algorithm for the problem S. Then the error of ϕ is given by

$$e(\phi,N) = \sup_{\| Tf \| \le 1} \| S(f) - \phi(N(f)) \|. \tag{E.13}$$

Let ϕ_1: $N(F_1) \to F_2$ be an algorithm for the problem S_1. Then the error of ϕ_1 is

$$e(\phi_1,N) = \sup_{f \in F_1} \| S(f) - \phi_1(N(f)) \| \, \rho(\| Tf \|). \tag{E.14}$$

Assume that ϕ is homogeneous, i.e., $\phi(N(cf)) = c\phi(N(f))$. Then ϕ can be extended to $N(F_1)$ in such a way that its extension is also homogeneous. This extension is defined as

$$\phi_1(N(f)) = \begin{cases} \phi(N(f)) & \text{if } \| Tf \| \le 1, \\ \| Tf \| \phi(N(f/\| Tf \|)) & \text{if } \| Tf \| > 1. \end{cases} \tag{E.15}$$

We prove

LEMMA E.2 Let ϕ be a homogeneous algorithm and let ϕ_1 be its extension defined by (E.15). If $e(\phi,N) < +\infty$, then

$$e(\phi_1,N) = \sup_{x>0} x\,\rho(x)\,e(\phi,N) \tag{E.16}$$

with the convention $0 \cdot \infty = 0$. ■

PROOF From (E.14) we have

$$e(\phi_1,N) = \sup_{x\geq 0} \rho(x)\Big(\sup_{\|Tf\|=x} \|\,S(f) - \phi_1(N(f))\,\|\Big).$$

Note that $e(\phi,N) < +\infty$ implies that $\phi(N(f)) = S(f)$, $\forall f \in \ker T$. Assume first that $T \equiv 0$. Then $e(\phi_1,N) = e(\phi,N) = 0$, which agrees with (E.16). Assume now that $T \not\equiv 0$. From this and homogeneity of ϕ_1 we get

$$e(\phi_1,N) = \sup_{x>0} x\,\rho(x)\Big(\sup_{\|T(\frac{1}{x}f)\|=1} \|\,S(\tfrac{1}{x}f) - \phi_1(N(\tfrac{1}{x}f))\,\|\Big).$$

From (E.15) and (E.13) we conclude that

$$e(\phi_1 N) = \sup_{x>0} x\,\rho(x)\Big(\sup_{\|Tf\|=1} \|\,S(f) - \phi(N(f))\,\|\Big) = \sup_{x>0} x\,\rho(x)\,e(\phi,N),$$

which proves (E.16). ■

From Theorem E.1 and Lemma E.2 we easily obtain

LEMMA E.3 Let $\sup\{\rho(x): x \geq 0\} < +\infty$ and let

$$r(N,S,T) = \tfrac{1}{2}d(N,S,T) < +\infty. \tag{E.17}$$

If there exists a homogeneous optimal error algorithm ϕ for the problem S, then its extension ϕ_1, defined by (E.15), is a homogeneous optimal error algorithm for the problem S_1 and

$$r(N,S_1,T) = \sup_{x>0} x\,\rho(x)\,r(N,S,T). \tag{E.18}$$

■

PROOF Let $a = \sup\{x\,\rho(x): x > 0\}$. Note that $r(N,S,T) = \frac{1}{2}d(N,S,T) < +\infty$ implies that $\ker T \cap \ker N \subset \ker S$. Then, using successively (E.11), (E.10), and (E.16), we get

$$a\, r(N,S,T) = a\, d(N,S,T)/2 = d(N,S_1,T)/2 \le r(N,S_1,T)$$

$$\le e(\phi_1,N) = a\, e(\phi,N) = a\, r(N,S,T).$$

Hence, (E.18) is proven. ■

The assumptions of Lemma E.3 are satisfied, for instance, for the Hilbert space setting. Namely, if F_4 is a Hilbert space, $N(f) = [L_1(f),L_2(f),\ldots,L_n(f)]$, for linear functions $L_1,L_2,\ldots,L_n$ and $T(\ker N)$ is closed, then there exist elements $g_1,g_2,\ldots,g_n$ of F_1 such that

$$\phi(N(f)) = \sum_{i=1}^{n} L_i(f)Sg_i \tag{E.19}$$

is a strongly optimal error algorithm and $r(N,S,T) = \frac{1}{2}d(N,S,T)$. Note that in this case the extension ϕ_1 of the algorithm ϕ is linear and has the same form (E.19). This property will be extensively used in the study of average case models.

We briefly discuss optimal information operators for the problem S_1. If the assumptions of Theorem E.1 are satisfied and $\sup_{x>0} x\,\rho(x)$ is finite, then (E.11) states that the diameters of information for the problem S_1 are proportional to the diameters of information for the problem S. Thus, if an information operator N_n with $\mathrm{card}(N_n) \le n$ has minimal diameter for the problem S, then the same information operator N_n has minimal diameter for the problem S_1. Since the diameter of information always differs from the radius by at most a factor of two, then an nth optimal information operator for the problem S is almost nth optimal for the problem S_1, and vice versa. If F_4 is a Hilbert space, then

$$r(N,S,T) = \tfrac{1}{2}d(N,S,T) \quad \text{and} \quad r(N,S_1,T) = \tfrac{1}{2}d(N,S_1,T), \quad \forall n.$$

This implies that optimal information operators for the problems S and S_1 coincide. We summarize this in

COROLLARY E.1 Optimal information operators for the problems S and S_1 are essentially the same.

If F_4 is a Hilbert space, then they are the same. ■

APPENDIX F

GENERALIZED MEASURE OF UNCERTAINTY

Throughout this monograph we have assumed, for simplicity, that the uncertainty is measured by a nonnegative real number ϵ. In this appendix we extend some of our results to the case where ϵ is an element of a partially ordered set. This, of course, includes the case where ϵ is a vector. Our motivation in generalizing ϵ is twofold:

1. Our model cannot include certain applications if uncertainty is measured by a real parameter.

2. Since our basic concepts F, G, S, N are abstract sets and operators, it is in the spirit of our work also to choose ϵ as an element of an abstract set.

Since our proofs used only the order and inf properties of real numbers, it is not surprising that many concepts and results can be generalized. For simplicity we restrict ourselves to the class of idealized algorithms and exact information operators. We illustrate the concepts presented in this appendix by the pointwise approximation problem studied in the next appendix.

Thus, let ϵ be an element of a partially ordered set Δ (with the relation $\le$). For simplicity we assume that for every subset Δ_1 of Δ, $\inf \Delta_1$ exists. Recall that by $\inf \Delta_1$ we mean a unique element of Δ such that

$$\inf \Delta_1 \le \Delta, \qquad \forall \delta \in \Delta_1,$$

and if there exists δ_1 from Δ such that $\delta_1 \le \delta$, $\forall \delta \in \Delta_1$, then $\delta_1 \le \inf \Delta_1$. Thus, $\inf \Delta_1$ is the largest lower bound of Δ_1. Let $0 = \inf \Delta$, that is, $0 \le \delta$, $\forall \delta \in \Delta$.

We proceed analogously to Chapter 1 and define the solution operator S as

$$S\colon F \times \Delta \to 2^G, \tag{F.1}$$

where F and G are two given sets. We assume that S has two properties

$$S(f,0) \neq \emptyset, \qquad \forall f \in F, \tag{F.2}$$

$$\delta_1 \leq \delta_2 \text{ implies } S(f,\delta_1) \subset S(f,\delta_2) \tag{F.3}$$

for all $\delta_1,\delta_2 \in \Delta$ and all $f \in F$. Note that (F.2) and (F.3) have exactly the same form as (P_s.1) and (P_s.2) of Chapter 1; however the relation $\leq$ may now have a different meaning.

Let $N: F \to H$ be an (exact) information operator. For $f \in F$ recall that

$$V(N,f) = \{\tilde{f} \in F: N(\tilde{f}) = N(f)\},$$

$$A(N,f,\delta) = \bigcap_{\tilde{f} \in V(N,f)} S(\tilde{f},\delta).$$

We define the *local radius of information* as

$$r(N,f) = \inf\{\delta \in \Delta: A(N,f,\delta) \neq \emptyset\} \tag{F.4}$$

and the (global) *radius of information* as

$$r(N) = \inf\{\delta \in \Delta: A(N,f,\delta) \neq \emptyset,\ \forall f \in F\} \tag{F.5}$$

(compare with (3.3) and (3.4) of Chapter 1). Note that $r(N,f) \leq r(N)$, $\forall f \in F$. Furthermore, if $A(N,f,r(N,f)) \neq \emptyset$, $\forall f \in F$, then, taking δ from Δ such that $r(N,f) \leq \delta$, $\forall f \in F$, we conclude that $A(N,f,\delta) \neq \emptyset$, $\forall f \in F$. Thus $r(N) \leq \delta$. This shows that

$$r(N) = \sup_{f \in F} r(N,f).$$

Let ϕ: $N(F) \to G$ be an (idealized) algorithm. We define the *local error of* ϕ as

$$e(\phi,N,f) = \inf\{\delta \in \Delta: \phi(N(f)) \in A(N,f,\delta)\} \tag{F.6}$$

and the (global) *error of* ϕ as

$$e(\phi,N) = \inf\{\delta \in \Delta: \phi(N(f)) \in A(N,f,\delta),\ \forall f \in F\} \tag{F.7}$$

(compare with (3.6) and (3.7) of Chapter 1). Once more note that $e(\phi,N,f) \leq e(\phi,N)$, $\forall f \in F$, and if $A(N,f,\ e(\phi,N,f)) \neq \emptyset$, $\forall f \in F$, then

$$e(\phi,N) = \sup\{e(\phi,N,f): f \in F\}.$$

We are ready to prove a generalization of Theorem 3.1 of Chapter 1.

THEOREM F.1

$$\inf_{\phi \in \Phi(N)} e(\phi,N,f) = r(N,f), \qquad \forall f \in F, \tag{F.8}$$

$$\inf_{\phi \in \Phi(N)} e(\phi,N) = r(N), \tag{F.9}$$

where $\Phi(N)$ is, as always, the class of all algorithms that use N. ■

PROOF First we show that $e(\phi,N,f) \geq r(N,f)$. Indeed, let $\phi(N(f)) \in A(N,f,\delta)$. Thus $A(N,f,\delta) \neq \emptyset$ and $r(N,f) \leq \delta$. Since this holds for all such δ, then $r(N,f) \leq e(\phi,N,f)$. Since ϕ is arbitrary, we have

$$r(N,f) \leq a \stackrel{df}{=} \inf\{e(\phi,N,f): \phi \in \Phi(N)\}.$$

Now take δ such that $A(N,f,\delta) \neq \emptyset$. Then there exists an element $g \in A(N,f,\delta)$. Defining $\phi(N(f)) = g$, we conclude that $e(\phi,N,f) \leq \delta$ and $a \leq \delta$. Since δ is arbitrary, $a \leq r(N,f)$, which proves that $a = r(N,f)$. Hence (F.8) holds. We omit the proof of (F.9) since it can be established exactly as (F.8) ■

Assume that

$$A(N,f,r(N,f)) \neq \emptyset, \ \forall f \in F. \tag{F.10}$$

An algorithm ϕ^c is *central* iff

$$\phi^c(N(f)) \in A(N,f,r(N,f)), \qquad \forall f \in F.$$

From Theorem F.1 and (F.6) we obtain a generalization of Theorem 4.1 of Chapter 1.

COROLLARY F.1 A central algorithm ϕ^c is a strongly optimal error algorithm, i.e.,

$$e(\phi^c,N,f) = r(N,f), \qquad \forall f \in F,$$

and

$$e(\phi^c,N) = r(N).$$ ■

We can now verify when it is possible to find an ϵ-approximation using the information operator N. By an ϵ-approximation we mean an element x such that $x \in S(f,\epsilon)$ for a given ϵ which belongs to the set Δ.

COROLLARY F.2 Let (F.10) hold. Then it is possible to find an ϵ-approximation $\forall f \in F$ iff $r(N) \leq \epsilon$. ■

PROOF Assume that there exists an algorithm ϕ that uses N such that $\phi(N(f))$ is an ϵ-approximation for all f from F, that is, $\phi(N(f)) \in S(f,\epsilon)$, $\forall f \in F$. Then $\phi(N(f)) \in A(N,f,\epsilon)$, $\forall f \in F$, which yields $e(\phi,N) \le \delta$. Due to (F.9), $r(N) \le \epsilon$.

Conversely, assume that $r(N) \le \epsilon$. Take a central algorithm ϕ^c. Then $\phi^c(N(f)) \in A(N,f,r(N,f))$. Since $r(N,f) \le r(N) \le \epsilon$, we have $\phi^c(N(f)) \in S(f,\epsilon)$, $\forall f \in F$. Thus $\phi^c(N(f))$ is an ϵ approximation, which completes the proof. ∎

We now briefly discuss the optimal information problem. In Chapter 4 we compare two information operators N_1 and N_2 of the same cardinality by their radii $r(N_1)$ and $r(N_2)$. We prefer to use N_1 if $r(N_1) \le r(N_2)$. This approach cannot be taken here because $r(N_1)$ and $r(N_2)$ are not, in general, comparable, i.e., neither $r(N_1) \le r(N_2)$ nor $r(N_2) \le r(N_1)$ holds.

Note also that although (2.8), (2.9), and (2.10) of Chapter 4 can be formally generalized, the nth minimal radius need not be attained by (or even be close to) the radius of some information operator of cardinality at most n. The reader can verify that the pointwise approximation problem studied in Appendix G is an example of such a case.

Since this property is undesirable, we do not work with nth optimal information. We choose to generalize the ϵ-cardinality number (see (2.11) of Chapter 4). Let $\Psi^{\text{non}}(n)$ and $\Psi^{a}(n)$ be defined as in Chapter 4. Then

$$\begin{aligned} m^{\text{non}}(\epsilon) &= \min\{n\colon N \in \Psi^{\text{non}}(\mathrm{n}),\ r(N) \le \epsilon\}, \\ m^{a}(\epsilon) &= \min\{n\colon N \in \Psi^{a}(n),\ r(N) \le \epsilon\}, \end{aligned} \tag{F.11}$$

is the *ϵ-cardinality number for nonadaptive (adaptive) information.*

Thus $m^{\text{non}}(\epsilon)$ is the minimal cardinality of any information operator from the class $\Psi^{\text{non}}(n)$ whose radius does not exceed ϵ. Modulo the assumption (F.10), we can also say that $m^{\text{non}}(\epsilon)$ is the minimal number of primitive evaluations to find an ϵ-approximation.

APPENDIX G

POINTWISE APPROXIMATION

As an application of partially ordered error criteria presented in Appendix F, we pose and solve questions concerning optimal information and optimal algorithms for pointwise approximation. Although this material belongs among the applications of Chapter 6, we present it here because it illustrates some of the theory of Appendix F.

The pointwise approximation problem is roughly the following. Given information $N(f)$ concerning a function $f(x)$, compute a function $g(x)$ that differs from $f(x)$ by at most $\epsilon(x)$. The fact that one may choose different error criteria at different x makes the approximation pointwise.

For simplicity we assume here that f is differentiable and $N(f)$ consists of function samples. These conditions could be modified and one could study pointwise approximation for the class of r times differentiable f or for $N(f)$ consisting of linear functionals.

We pose two problems:

1. What is the minimal number of function evaluations required to compute an ϵ-approximation? What is the optimal choice of the points of evaluation?

2. Obtain an algorithm with close to optimal complexity that uses this minimal number of function evaluations to compute an ϵ-approximation.

We completely solve the first problem and present an elegant geometric method for determining the optimal evaluation points. We solve the second problem by obtaining an optimal error algorithm that uses the optimal information and enjoys linear combinatory complexity. This algorithm must therefore have been close to optimal complexity.

For particular choices of the uncertainty function $\epsilon(x)$, we obtain "absolute" and "relative" error criteria. We prove that the absolute criterion requires, in terms of the interval length, exponentially more function evaluations than the relative criterion.

Let $\Delta = \{\delta\colon\ \delta:[0,b] \to \mathbb{R}_+\}$, $b > 0$, be the set of nonnegative functions. By

$\delta_1 \le \delta_2$ we mean $\delta_1(x) \le \delta_2(x)$, $\forall x \in [0,b]$. Let Δ_1 be a subset of Δ. Then inf Δ_1 is a function such that

$$(\inf \Delta_1)(x) = \inf\{\delta(x): \delta \in \Delta_1\}, \qquad \forall x \in [0,b], \tag{G.1}$$

where the second "inf" denotes the usual infimum of the corresponding set of nonnegative numbers.

Let

$$F = G = \{f: f:[0,b] \to \mathbb{R}, f \text{ is absolutely continuous}, \mid f'(t) \mid \le 1 \text{ a.e.}\}. \tag{G.2}$$

For $\epsilon \in \Delta$ and $f \in F$ define the solution operator as

$$S(f,\epsilon) = \{g \in G: \mid f(x) - g(x) \mid \le \epsilon(x), \forall x \in [0,b]\}. \tag{G.3}$$

Thus, to find an ϵ-approximation means to approximate the function f in such a way that the difference at each x of $[0,b]$ does not exceed the value $\epsilon(x)$. That's why we call this problem the *pointwise approximation problem.*

Remark G.1: Choices of ϵ Let g be an ϵ-approximation. Then $f(x) = g(x)$ whenever $\epsilon(x) = 0$. Thus g interpolates f at all points at which ϵ vanishes. If $\epsilon(x) \equiv \epsilon$, then $\| f - g \| \le \epsilon$, where $\| f \| = \sup\{\mid f(x) \mid : x \in [0,b]\}$. In this case, an ϵ-approximation means that the absolute error does not exceed ϵ. If $\epsilon(x) = a_1 x + a_2$ with (presumably small) nonnegative a_1 and a_2, then the absolute error $\mid f(x) - g(x) \mid$ is smaller than a_2 for small x, and the relative error $\mid f(x) - g(x) \mid /x$ is smaller than a_1 for large x. In this case ϵ is a combination of absolute and relative errors. ■

Remark G.2: Relation to Uniform Approximation The pointwise approximation problem generalizes the usual approximation problem where we seek a function g such that $\| f - g \| \le \epsilon$ for some norm and a nonnegative number ϵ.

Let $d(f) = \sup\{\mid f(x) \mid / \epsilon(x): x \in [0,b]\}$ with the convention $0/0 = 0$. Then g is an ϵ-approximation iff $d(f - g) \le 1$. If $\min_{x\in[0,b]} \epsilon(x) > 0$, then d is the supremum norm with the weight function $1/\epsilon$. In this case the pointwise approximation problem corresponds to the uniform weighted approximation problem. On the other hand, if $\min_{x\in[0,b]} \epsilon(x) = 0$, then d is not well defined for some f from F and the pointwise approximation problem is not related to the uniform weighted approximation problem. ■

Let

$$N(f) = [f(x_n), f(x_{n-1}),\ldots, f(x_1)] \tag{G.4}$$

be an information operator, where $0 \le x_n < x_{n-1} < \cdots < x_1 \le b$. Note that N is a nonadaptive information operator.

We obtain the central algorithm, and the local and global radii of information. Let

$$w_i = \frac{x_i + x_{i+1}}{2} + \frac{f(x_{i+1}) - f(x_i)}{2},$$

$$z_i = \frac{x_i + x_{i+1}}{2} - \frac{f(x_{i+1}) - f(x_i)}{2},$$

for $i = 1,2, \ldots, n-1$. Define

$$f^-(x) = \begin{cases} f(x_n) + x - x_n & \text{if } 0 \le x \le x_n, \\ f(x_{i+1}) - x + x_{i+1} & \text{if } x_{i+1} \le x \le w_i, \\ f(x_i) + x - x_i & \text{if } w_i \le x \le x_i, \\ f(x_1) - x + x_1 & \text{if } x_1 \le x \le b, \end{cases}$$

$$f^+(x) = \begin{cases} f(x_n) - x + x_n & \text{if } 0 \le x \le x_n, \\ f(x_{i+1}) + x - x_{i+1} & \text{if } x_{i+1} \le x \le z_i, \\ f(x_i) - x + x_i & \text{if } z_i \le x \le x_i, \\ f(x_1) + x - x_1 & \text{if } x_1 \le x \le b. \end{cases}$$

Then, as in the integration example, we have

$$f^-(x) \le \tilde{f}(x) \le f^+(x), \qquad \forall x \in [0,b], \qquad \forall \tilde{f} \in V(N,f). \tag{G.5}$$

Note that f^- and f^+ belong to $V(N,f)$. Their graphs are presented in Fig. G.1.

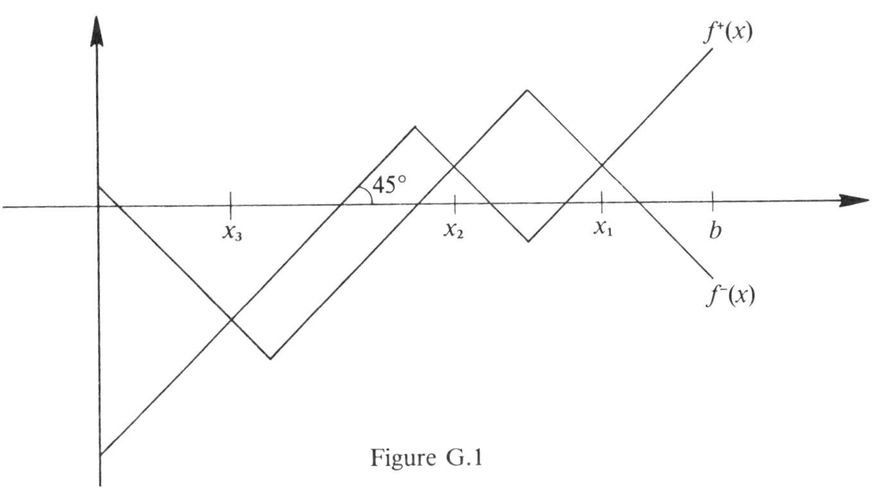

Figure G.1

Define the algorithm

$$\phi^*(N(f))(x) = \tfrac{1}{2}[f^+(x) + f^-(x)]. \tag{G.6}$$

Note that the complexity of the computation of $\phi^*(N(f))(x)$ is proportional to the cost of n function evaluations and n arithmetic operations. Thus, the complexity of ϕ^* is linear in n. From (G.5) we have

$$|\tilde{f}(x) - \phi^*(N(f))(x)| \le \epsilon^*(x;N(f)) \overset{df}{=} \tfrac{1}{2}[f^+(x) - f^-(x)], \quad \forall \tilde{f} \in V(N,f). \tag{G.7}$$

The graph of $\epsilon^*(\cdot;N(f))$ is presented in Fig. G.2.

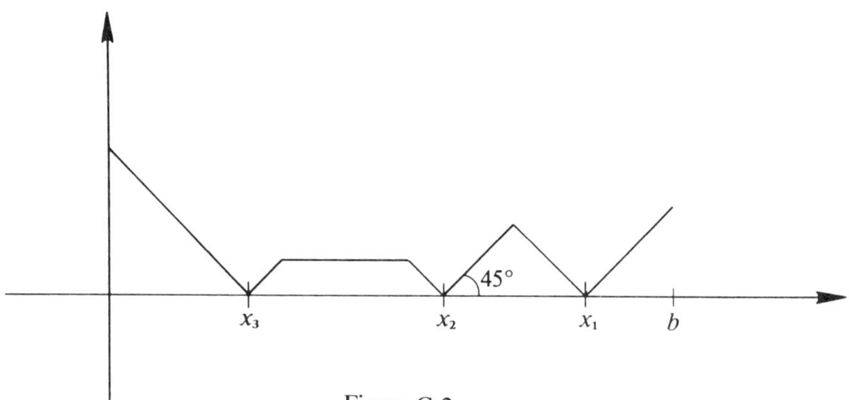

Figure G.2

We prove

LEMMA G.1 The algorithm ϕ^* is central and

$$r(N,f) = e(\phi^*,N,f) = \epsilon^*(\cdot;N(f)), \tag{G.8}$$

$$r(N) = r(N,0) = \epsilon^*(\cdot;0). \tag{G.9}$$

PROOF We first show that $r(N,f) = e(\phi^*,N,f) = \epsilon^*(\cdot;N(f))$. Take δ such that $A(N,f,\delta) \ne \emptyset$. Let $g \in A(N,f,\delta)$. Since $f^-,f^+ \in V(N,f)$,

$$|f^-(x) - g(x)| \le \delta(x) \quad \text{and} \quad |f^+(x) - g(x)| \le \delta(x), \quad \forall x \in [0,b].$$

Thus

$$\epsilon^*(x,N(f)) = \tfrac{1}{2}\{f^+(x) - g(x) - [f^-(x) - g(x)]\} \le \delta(x), \qquad \forall x \in [0,b].$$

Since δ is arbitrary, $\epsilon^*(\cdot;N(f)) \le r(N,f)$. Observe that (G.7) yields

$$\phi^*(N(f)) \in A(N,f,\epsilon^*(\cdot;N(f)).$$

Thus $e(\phi^*,N,f) \le \epsilon^*(\cdot;N(f))$. Due to (F.8), we have $r(N,f) = e(\phi^*,N,f) = \epsilon^*(\cdot;N(f))$, which proves (G.8) and that ϕ^* is central. To prove (G.9) it is sufficient to observe that $\epsilon^*(\cdot;N(f)) \le \epsilon^*(\cdot;0)$. ■

The algorithm ϕ^* is central and its complexity is linear in n. Thus, ϕ^* is a nearly optimal complexity algorithm for N.

Note that (F.10) holds for our example. Thus Corollary F.1 and Lemma F.1 yield that we can find an ϵ-approximation iff

$$\epsilon^*(x;0) \le \epsilon(x), \qquad \forall x \in [0,b]. \tag{G.10}$$

The graph of $\epsilon^*(\cdot;0)$ is presented in Fig. G.3.

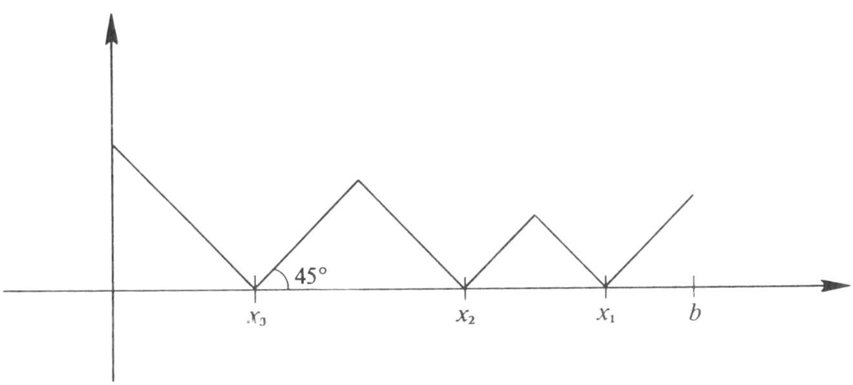

Figure G.3

We now find the ϵ-cardinality number, i.e., the minimal number of points x_i such that $\epsilon^*(x;0) \le \epsilon(x)$, $\forall x \in [0,b]$. It will be convenient to extend the domain of ϵ by setting $\epsilon(x) = \epsilon(0)$ for $x \le 0$. For simplicity we also assume that ϵ is continuous.

We define two sequences $\{x_i\}$ and $\{y_i\}$ by induction. Let $y_1 = b$. Assume that y_i is already defined, $i \ge 1$. Then let

$$x_i = \max\{x - \epsilon(x) : x \in (-\infty,y_i)\} \tag{G.11}$$

and let y_{i+1} be the smallest number such that

$$\epsilon(x) \ge x_i - x \qquad \text{for} \qquad x \in [y_{i+1},x_i]. \tag{G.12}$$

We illustrate the definition of $\{x_i\}$ and $\{y_i\}$ by Fig. G.4.

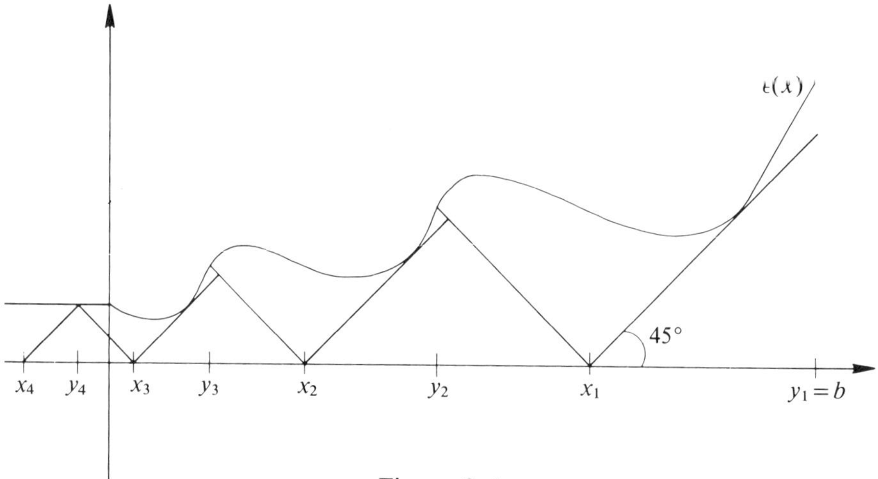

Figure G.4

We describe the construction of $\{x_i\}$ and $\{y_i\}$ in words. The point x_1 is constructed by taking a line $x - x_1$ with maximal x_1 such that it meets $\epsilon(x)$. (Thus $\epsilon(x) \geq x - x_1$ and $\epsilon(x) > x - a$ for $a > x_1$, and consequently $x_1 = \max\{x - \epsilon(x): x \leq b\}$.) To construct y_2, we draw the line $x_1 - x$ and compare with $\epsilon(x)$. The point y_2 is the smallest number such that $\epsilon(x) \geq x_1 - x$ for $x \in [y_2, x_1]$. Next we replace b by y_2 and repeat the above construction getting x_2 and y_3, and so on.

Observe that continuity of ϵ implies that x_i and y_{i+1} are well defined. Note that

$$x_{i-1} \leq y_{i+1} \leq x_i \leq y_i \leq b, \qquad \forall i,$$

and x_i is the smallest number such that

$$\epsilon(x) \geq x - x_i \qquad \text{for} \qquad x \in (-\infty, y_i),$$

whereas y_{i+1} is a zero of the equation $\epsilon(x) = x_i - x$. If $\epsilon(x_i) = 0$ and if the left-sided $\epsilon'(x_i)$ exists and is larger than -1, then $y_{i+1} = x_i$.

From (G.11) we have $x_i \geq y_i - \epsilon(y_i)$; from (G.12) it follows that $\epsilon(y_i) = x_{i-1} - y_i$. Thus $y_i \leq x_i + \epsilon(y_i) = x_i + x_{i-1} - y_i$. Hence

$$y_i \leq (x_i + x_{i-1})/2. \tag{G.13}$$

Observe also that

$$y_i = x_i \quad \text{implies} \quad x_{i-1} = x_i. \tag{G.14}$$

Indeed, $y_i = x_i$, and (G.11) yield $\epsilon(y_i) = 0$. Since $\epsilon(x) \geq x_{i-1} - x$ for $x \in [y_i, x_{i-1}]$ due to (G.12), then, setting $x = y_i$, we get $0 \geq x_{i-1} - y_i$. Thus $x_{i-1} = y_i = x_i$. Let

$$n^* = n(\epsilon) = \max\{i: y_i > 0\} \tag{G.15}$$

with the convention that $n^* = n(\epsilon) = +\infty$ if $y_i > 0$, $\forall i$. We stress that n^* may be equal to infinity. Indeed, if $\epsilon \equiv 0$, then (G.11) and (G.12) yield $y_i = x_i = b$, $\forall i$. Note also that $n^* = +\infty$ if there exists a nonnegative x^* less than b such that $\epsilon(x^*) = 0$ and $\epsilon'(x^*) < 1$.

Suppose for a moment that n^* is finite. Define the information operator

$$N_{n^*}(f) = [f(\max(x_{n^*},0)), f(x_{n^*-1}), \cdots, f(x_1)]. \tag{G.16}$$

Note that $x_{n^*-1} \geq y_{n^*} > 0$, which means that we compute f at nonnegative points. Since all x_i are given independently of f, N_{n^*} is a nonadaptive information operator of cardinality n^*. We prove

LEMMA G.2 If $n^* = n(\epsilon)$ is finite, then $r(N_{n^*}) \leq \epsilon$. ■

PROOF Let $x \in [x_1, b]$. Then Lemma G.1 and (G.11) yield $r(N_{n^*})(x) = x - x_1$ and $x_1 \geq x - \epsilon(x)$. Thus $r(N_{n^*})(x) \leq \epsilon(x)$. Let $x \in [x_i, x_{i-1}]$ with $x_i > 0$. Then $y_i > 0$ and $i \leq n^*$. If $x \leq y_i$, then, due to (G.13), $x \leq (x_i + x_{i-1})/2$ and $r(N_{n^*})(x) = x - x_i$. Due to (G.11), $x_i \geq x - \epsilon(x)$. Thus $r(N_{n^*})(x) \leq \epsilon(x)$. If $x \in [y_i, x_{i-1}]$, then $r(N_{n^*})(x) \leq x_{i-1} - x$. Since $\epsilon(x) \geq x_{i-1} - x$, due to (G.12) $r(N_{n^*})(x) \leq \epsilon(x)$.

Consider now the interval $[x_{i+1}, x_i]$ with $x_{i+1} \leq 0$ and $x_i > 0$. Assume first that $y_{i+1} \leq 0$. Then $i = n^*$. For $x \in [0, x_{n^*}]$ we have $r(N_{n^*})(x) = x_{n^*} - x \leq \epsilon(x)$ due to (G.12). Assume then that $y_{i+1} > 0$. Then $n^* = i + 1$. Since we know $f(0)$, $r(N_{n^*})(x) = x$ for $x \in [0, y_{n^*}]$. Thus $r(N_{n^*})(x) \leq x - x_{n^*} \leq \epsilon(x)$ due to (G.11). If $x \in [y_{n^*}, x_{n^*-1}]$, then $r(N_{n^*})(x) \leq x_{n^*-1} - x \leq \epsilon(x)$ due to (G.12). This completes the proof. ■

We now show that $n(\epsilon)$ is the ϵ-cardinality number. More precisely, let $\Psi^{\text{non}}(n)$ be the class of nonadaptive information operators of the form

$$N(f) = [f(a_1), f(a_2), \ldots, f(a_n)]$$

for some points $a_1, a_2, \ldots, a_n$ from $[0,b]$. Let $\Psi^a(n)$ denote the same class with adaptively chosen $a_1, a_2, \ldots, a_n$. Now we are ready to prove the main result of this appendix.

THEOREM G.1 The ϵ-cardinality number is given by

$$m^{\text{non}}(\epsilon) = m^a(\epsilon) = n(\epsilon). \tag{G.17}$$

■

PROOF Let $N \in \Psi^a(n)$ with $r(N) \le \epsilon$. Thus $N(f) = [f(a_1),\dots,f(a_n)]$, where

$$a_2 = a_2(f(a_1)), \ \dots, \ a_n = a_n(f(a_1), \ \dots, \ f(a_{n-1})).$$

For $f \equiv 0$, we compute f at the points a_1, $a_2(0)$, ..., $a_n(0,\dots,0)$. Let $a_1^*, a_2^*,\dots,a_n^*$ be a permutation of $a_1, a_2(0), \dots, a_n(0,\dots,0)$ such that $b \ge a_1^* \ge a_2^* \ge \cdots \ge a_n^* \ge 0$.

We show that a_1^* belongs to $[x_1,b]$. Suppose, on the contrary, that $a_1^* < x_1$. It follows from (G.11) that $x_1 = x^* - \epsilon(x^*)$ for some x^* from $(-\infty,b)$. Since $\epsilon(x^*) \ge 0$, then $x_1 \le x^*$. From (G.8) of Lemma G.1 we know that $r(N,0)(x^*) = x^* - a_1^*$. Thus $r(N,0)(x^*) > x^* - x_1 = \epsilon(x^*)$, which contradicts $r(N) \le \epsilon$. Hence $a_1^* \in [x_1,b]$.

We now show that for every subinterval $[x_i,x_{i-1}]$ with $x_i \ge 0$ there exists a_j^* such that $a_j^* \in [x_i,x_{i-1}]$. Suppose, on the contrary, that $a_j^* < x_i$ and $x_{i-1} < a_{j-1}^*$ for some j (with the convention $a_{n+1}^* = -\infty$). We know that $x_i = x^* - \epsilon(x^*)$, where $x^* \in [x_i,y_i] \subset [x_i,x_{i-1}]$. From (G.8) we conclude that $r(N,0)(x^*) = \min(x^* - a_j^*, a_{j-1}^* - x^*)$. Note that $x^* - a_j^* > x^* - x_i = \epsilon(x^*)$. Due to (G.13), we have $x^* \le y_i \le (x_i + x_{i-1})/2$. Thus $x_{i-1} - x^* \ge x^* - x_i$ and $a_{j-1}^* - x^* > x_{i-1} - x^* \ge x^* - x_i = \epsilon(x^*)$. So $r(N,0)(x^*) > \epsilon(x^*)$, which is a contradiction. Hence there exists a_j^* such that $a_j^* \in [x_i,x_{i-1}]$.

Let $k = \sup\{i\colon b = x_0 > x_1 > x_2 > \cdots > x_i \ge 0\}$ be the total number of distinct subintervals $[x_i,x_{i-1}]$ of $[0,b]$. Note that k may be infinity! Since each $[x_i,x_{i-1}]$ contains at least one element of the sequence $a_1^*,a_2^*,\dots,a_n^*$, we conclude that $n \ge k$.

Assume that $k = +\infty$. Then we have infinitely many positive y_i and $n(\epsilon) = +\infty$. Since $r(N) \le \epsilon$ implies $n = +\infty$, there exists no adaptive information operator of finite cardinality with radius less than or equal to ϵ. Thus $m^{\text{non}}(\epsilon) = m^a(\epsilon) = +\infty$, which proves (G.17) in this case.

Assume then that k is finite. Consider the point x_{k+1}. Then $x_{k+1} < 0$ or $x_{k+1} = x_k$. Suppose first that $x_{k+1} < 0$. From (G.14) we conclude that $y_k > 0$ and

$$k = \begin{cases} n(\epsilon) & \text{if } \ y_{k+1} \le 0, \\ n(\epsilon) - 1 & \text{if } \ y_{k+1} > 0. \end{cases}$$

If $y_{k+1} \le 0$, then $n \ge k = n(\epsilon)$. Since N is arbitrary, we get $m^a(\epsilon) \ge n(\epsilon)$.

If $y_{k+1} > 0$, we show that $a_n^* \in [0,x_k]$. Suppose, on the contrary, that $a_n^* > x_k$. Then $r(N,0)(y_{k+1}) = a_n^* - y_{k+1} > x_k - y_{k+1} = \epsilon(y_{k+1})$, which is a contradiction. Thus $[0,x_k]$ contains a_n^*, showing that $n \ge k + 1 = n(\epsilon)$. Hence $m^a(\epsilon) \ge n(\epsilon)$ in this case also.

Suppose then that $x_{k+1} = x_k$. Then $y_{k+1} = x_k$ and, consequently, $x_{k+i} = y_{k+i} = x_k$, $\forall i \ge 1$. From (G.12) it follows that $\epsilon(x_k) = 0$ and there exists a positive δ_1 such that $\epsilon(x) < x_k - x$ for $x \in [x_k - \delta_1,x_k)$.

Assume that $x_k > 0$. Then $n(\epsilon) = +\infty$. We show that in this case $r(N)$

cannot be less than or equal to ϵ. Indeed, if $r(N,0) \leq \epsilon$, then $r(N,0)(x_k) = 0$, and from Lemma G.1 it follows that there exists a positive δ_2 such that $r(N,0)(x) = x_k - x$ for $x \in [x_k - \delta_2, x_k]$. Taking x from $[x_k - \min(\delta_1,\delta_2), x_k)$, we conclude that $r(N,0)(x) > \epsilon(x)$. Thus

$$m^{\text{non}}(\epsilon) = m^a(\epsilon) = n(\epsilon) = +\infty.$$

Now, assume that $x_k = 0$. Then $k = n(\epsilon)$, due to (G.14), and $n \geq n(\epsilon)$. Since N is arbitrary, $m^a(\epsilon) \geq n(\epsilon)$.

Thus we have proven (G.17) for $n(\epsilon) = +\infty$ and $m^a(\epsilon) \geq n(\epsilon)$ for a finite $n(\epsilon)$. From Lemma G.2 we know that $m^{\text{non}}(\epsilon) \leq n(\epsilon)$, which completes the proof. ■

Theorem G.1 states that $n(\epsilon)$ function evaluations are necessary to find an ϵ-approximation. If $n(\epsilon)$ is finite, the central and nearly optimal complexity algorithm ϕ^*, using the information (G.16), computes an ϵ-approximation. We stress that the information operator (G.16) is nonadaptive. Hence adaption does not help for the pointwise approximation problem. We summarize this in

COROLLARY G.1

(i) The minimal number of function evaluations necessary to find an ϵ-approximation is $n(\epsilon)$.

(ii) If $n(\epsilon)$ is finite, then the algorithm (G.6), which uses the nonadaptive information operator (G.16), yields an ϵ-approximation at a cost proportional to $n(\epsilon)$ function evaluations and $n(\epsilon)$ arithmetic operations. This algorithm is a central and nearly optimal complexity algorithm for the class $\Psi^a(n)$.

(iii) Adaption does not help in solving the pointwise approximation problem. ■

We now illustrate the construction of points x_i and the ϵ-cardinality number $n(\epsilon)$ by two examples.

Example G.1 Let $\epsilon(x) \equiv \epsilon$, $\epsilon > 0$. In this case, g is an ϵ-approximation if $\| f - g \| \leq \epsilon$, where $\| f \| = \sup_{x \in [0,b]} | f(x) |$. Then (G.11) and (G.12) yield

$$x_i = y_i - \epsilon, \qquad y_{i+1} = x_i - \epsilon.$$

Thus

$$x_i = b - (2i - 1)\epsilon, \qquad y_i = b - (2i - 2)\epsilon, \qquad \forall i \geq 1.$$

From this we have

$$n(\epsilon) = \left\lceil \frac{b}{2\epsilon} \right\rceil \simeq \frac{b}{2\epsilon} . \tag{G.18}$$

Note that $n(\epsilon)$ depends linearly on the length of the interval $[0,b]$. ■

Example G.2 Let $\epsilon(x) = a_1x + a_2$ with (small) positive a_1 and a_2. Now, g is an ϵ-approximation if

$$| f(x) - g(x) | \leq a_1x + a_2, \qquad \forall x \in [0,b].$$

We assume that $a_1 < 1$. From (G.11) and (G.12) we get

$$x_i = (1 - a_1)y_i - a_2 \qquad \text{and} \qquad (1 + a_1)y_{i+1} = x_i - a_2.$$

Let $q = (1 - a_1)/(1 + a_1)$. Then

$$x_i = qx_{i-1} - \frac{2a_2}{1 + a_1} = q^{i-1}x_1 - \frac{a_2}{a_1}(1 - q^{i-1}),\ i \geq 2,$$

$$x_1 = (1 - a_1)b - a_2.$$

Similarly we can show that

$$y_i = q^{i-1}b - \frac{a_2}{a_1}(1 - q^{i-1}).$$

From this we conclude that

$$n(\epsilon) = \left\lceil \frac{\log[a_2/(a_1b + a_2)]}{\log q} \right\rceil . \tag{G.19}$$

We analyze (G.19) for different values of a_1, a_2, and b.

(i) For fixed a_2 and b, let a_1 tend to zero. Then $\epsilon(x) \simeq a_2$ and $n(\epsilon) \simeq b/(2a_2)$, which agrees with (G.18) for $\epsilon = a_2$.

(ii) For fixed a_1 and b, let a_2 tend to zero. Then $\epsilon(x) \simeq a_1x$ and $n(\epsilon) \to +\infty$. This agrees with the remark below (G.15) with $x^* = 0$ that the ϵ-cardinality number is infinity if $\epsilon(0) = 0$ and $\epsilon'(0) < 1$.

(iii) For fixed a_1 and a_2, let b tend to infinity. Then $\epsilon(x) = a_1x + a_2$ and

$$n(\epsilon) \simeq \frac{\log b}{\log(1/q)} .$$

This shows that the total number of function evalutions depends on the logarithm of the interval length. This should be contrasted with (G.18) where the total number of function evaluations depends linearly on the interval length. ■

APPENDIX H

FURTHER LOWER BOUND RESULTS

In this appendix we exhibit further lower bounds on the local radii and the errors of algorithms. We study how a set $R = R(N_\rho)$, which defines the class of realizable algorithms, influences the radius of approximate information. We define the concept of a "distance" between $S(f,\eta)$ and $Q(N_\rho(f))$ and prove that it is a lower bound on the radius of approximate information.

Let N_ρ be an approximate information operator and let $R = R(N_\rho)$ be a class of realizable algorithms. Recall that

$$r(R,N_\rho,f) = \inf\{\delta: \bigcap_{\tilde{f}\in V(N_\rho,f)} S(\tilde{f},\delta) \cap Q(N_\rho(f)) \neq \emptyset\}$$

is the local radius. For $\delta > r(R,N_\rho,f)$, we know that $S(\tilde{f},\delta) \cap Q(N_\rho(f))$ is nonempty. The parameter δ measures the distance between the sets $S(\tilde{f},\delta)$ and $Q(N_\rho(f))$.

Let

$$\text{dist}(R,\tilde{f},f) = \inf\{\eta: S(\tilde{f},\eta) \cap Q(N_\rho(f)) \neq \emptyset\} \tag{H.1}$$

be the *local distance*. Let

$$\text{dist}(R,N_\rho,f) = \sup_{\tilde{f}\in V(N_\rho,f)} \text{dist}(R,\tilde{f},f) \tag{H.2}$$

$$(= \inf\{\eta: S(\tilde{f},\eta) \cap Q(N_\rho(f)) \neq \emptyset,\ \forall \tilde{f} \in V(N_\rho,f)\})$$

be the (global) *distance*.

Note that in the normed linear space setting

$$\text{dist}(R,\tilde{f},f) = \inf\{\| S(\tilde{f}) - x \| : x \in Q(N_\rho(f))\}$$

is the usual distance between $S(\tilde{f})$ and the set $Q(N_\rho(f))$. This motivates why $\text{dist}(R,\tilde{f},f)$ and $\text{dist}(R,N_\rho,f)$ are called distances in general. The concept of distance is schematized in Fig. H.1, where f_1 and f_2 are elements from $V(N_\rho,f)$ and $\text{dist}_i = \text{dist}(R,f_i,f)$, $i = 1,2$.

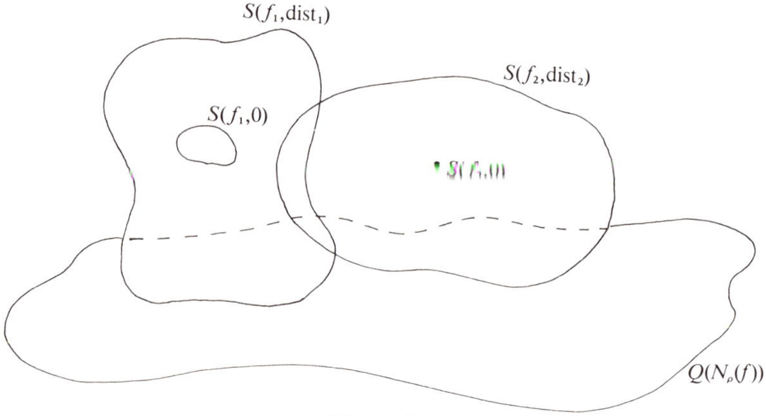

Figure H.1

LEMMA H.1

$$r(R,N_\rho,f) \geq \max(r(N_\rho,f),\ \text{dist}(R,N_\rho,f)). \tag{H.3}$$

■

PROOF Since $r(R,N_\rho,f) \geq r(N_\rho,f)$ for every R, it is sufficient to show that $r(R,N_\rho,f) \geq \text{dist}(R,N_\rho,f)$. Indeed, if $\eta < \text{dist}(R,N_\rho,f)$, then (H.2) yields that there exists $\tilde{f} \in V(N_\rho,f)$ such that $S(\tilde{f},\eta) \cap Q(N_\rho(f))$ is empty. Thus $\eta \leq r(R,N_\rho,f)$, which yields $\text{dist}(R,N_\rho,f) \leq r(R,N_\rho,f)$. Hence (H.3) is proven. ■

Lemma H.1 states that the local radius of approximate information (for R) is bounded from below by the local radius of approximate information and the global distance. For the normed linear space setting, Lemma 3.2 of Chapter 3 yields that (for $\rho = 0$)

$$r(R,N,f) \leq 2r(N,f) + \text{dist}(R,N,f). \tag{H.4}$$

We shall prove that, in general, it is impossible to bound the local radius $r(R,N_\rho,f)$ from above by $r(N_\rho,f)$ and $\text{dist}(R,N_\rho,f)$. We shall do this by describing a case where $r(R,N_\rho,f) = +\infty$ although both $r(N_\rho,f)$ and $\text{dist}(R,N_\rho,f)$ are finite.

LEMMA H.2 There exists a problem such that both $r(N_\rho,f)$ and $\text{dist}(R,N_\rho,f)$ are finite and $r(R,N_\rho,f) = +\infty$. ■

PROOF Consider the problem presented in the proof of Theorem A.1 with $d = +\infty$. Let $F = \{0,2\}$ and $R = \{\phi: \phi(h) \in F,\ \forall h\}$. Then $Q(h) = F$. Since $f \in S(f,\eta)$, $\forall \eta \geq 0$, then $S(f,\eta) \cap Q(N_\rho(f)) \neq \emptyset$. This implies that $\text{dist}(R,N_\rho,f) = 0$. Recall that $r(N_\rho,f) = 1$. Note that

$$S(0,\delta) \cap S(2,\delta) \cap Q(N_\rho(f)) \subset [-1,+1] \cap [1,3] \cap \{0,2\} = \emptyset.$$

Thus, $r(R,N_\rho,f) = +\infty$, which completes the proof. ■

There are two cases for which we have equality in (H.3). If $R = \Phi(N)$, then $\text{dist}(R,N_\rho,f) = 0$ and $r(R,N_\rho,f) = r(N_\rho,f)$. If N_ρ is complete, i.e., $V(N_\rho,f)$ is a singleton and $r(N_\rho,f) = 0$, then $r(R,N_\rho,f) = \text{dist}(R,N_\rho,f)$. We summarize this in

COROLLARY H.1 If $R = \Phi(N_\rho)$ or if N_ρ is complete, then

$$r(R,N_\rho,f) = \max(r(N_\rho,f),\ \text{dist}(R,N_\rho,f)). \tag{H.5}$$

■

We now turn to lower bounds on the error of a realizable algorithm ϕ that uses approximate information N_ρ. Let $y \in \phi(N_\rho(F))$. Let

$$\phi^{-1}(y) = \{N_\rho(f): f \in F \text{ and } \phi(N_\rho(f)) = y\} \tag{H.6}$$

be the pre-image of y. Let

$$W(\phi,y) = \bigcup_{N_\rho(f)\in\phi^{-1}(y)} V(N_\rho,f) \tag{H.7}$$

be the set of all elements f such that y is δ-approximation for $S(f,\delta)$ with $\delta > e(\phi)$. We illustrate (H.6) and (H.7) by Fig. H.2, where $i = 1, 2$ and $N_\rho(f_i) \in \phi^{-1}(y)$.

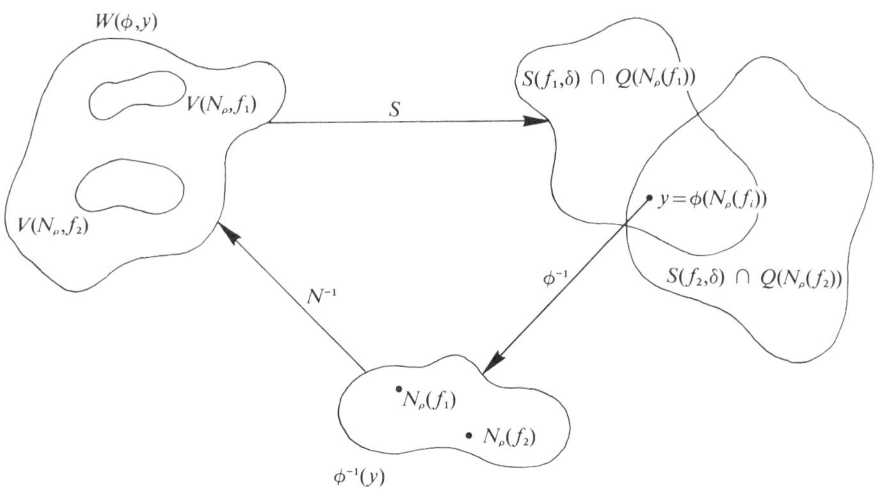

Figure H.2

We bound $e(\phi)$ from below as follows. Similarly to (H.1) and (H.2), define

$$\text{dist}(R,\phi,y) = \inf\{\eta: S(f_1,\eta) \cap Q(N_\rho(f_1)) \neq \emptyset, \forall f_1 \in W(\phi,y)\} \tag{H.8}$$

as the *local distance* of ϕ for y and

$$\text{dist}(R,\phi) = \sup\{\text{dist}(R,\phi,y): y = \phi(N_\rho(f)), f \in F\} \tag{H.9}$$

as the (global) *distance* of ϕ.

LEMMA H.3 Let $\phi \in R = R(N_\rho)$. Then

$$e(\phi) \geq \max(r(N_\rho), \text{dist}(R,\phi)). \tag{H.10}$$

■

PROOF Since $e(\phi) \geq r(R,N_\rho) \geq r(N_\rho)$, we need to show only that $e(\phi) \geq \text{dist}(R,\phi)$. Let $\eta < \text{dist}(R,\phi)$. Then there exists $y = \phi(N_\rho(f))$ with $f \in F$ such that $S(f_1,\eta) \cap Q(N_\rho(f_1)) \neq \emptyset$ for some $f_1 \in W(\phi,y)$. This implies that $e(\phi,f_1) \geq \eta$. Since $e(\phi) \geq e(\phi,f_1)$ and η is arbitrarily close to $\text{dist}(R,\phi)$, we get $e(\phi) \geq \text{dist}(R,\phi)$. Hence (H.10) is proven. ■

As before, we show that, in general, $e(\phi)$ cannot be bounded from above by $r(N_\rho)$ and $\text{dist}(R,\phi)$.

LEMMA H.4 There exists an algorithm $\phi \in R = R(N_\rho)$ such that $e(\phi) = +\infty$ although both $r(N_\rho)$ and $\text{dist}(R,\phi)$ are finite. ■

PROOF Consider the problem presented in the proof of Lemma H.2. Let ϕ be an interpolatory algorithm. Then $e(\phi) = +\infty$, due to Theorem A.1. Note that $W(\phi,y) = F$, and since $Q(h) = F$, we have $S(f,\delta) \cap F = \{f\}$, $\forall \delta \geq 0$. This yields $\text{dist}(R,\phi) = 0$. Since $r(N_\rho) = 1$, Lemma H.4 is proven. ■

Assume that ϕ is a central algorithm from R. Then equality in (H.10) holds for two cases. The first case is $R = \Phi(N_\rho)$ since then $\text{dist}(R,\phi) = 0$ and $e(\phi) = r(N_\rho)$. The second case is that N_ρ is complete. Then $r(N_\rho) = 0$. Since ϕ is central, we have

$$e(\phi,f) = \inf\{\delta: S(f,\delta) \cap Q(N_\rho(f)) \neq \emptyset\} \leq \text{dist}(R, \phi,y)$$

with $y = N_\rho(f)$ due to $f \in W(\phi,y)$. This and (H.10) yield $e(\phi) = \text{dist}(R,\phi)$. We summarize this in

COROLLARY H.2 Let ϕ be a central algorithm, $\phi \in R = R(N_\rho)$. If $R = \Phi(N_\rho)$ or N_ρ is complete, then $e(\phi) = \max(r(N_\rho), \text{dist}(R,\phi))$. ■

BIBLIOGRAPHY

Abelson, H. [78] Towards a theory of local and global in computation, *J. Theoret. Comptr. Sci.* **6** (1978), 41–67.

[80] Lower bounds on information transfer in distributed computations, *J. Assoc. Comput. Mach.* **27** (1980), 384–392.

Achieser, N.I. [56] *Theory of Approximation,* Frederich Ungar Publishing Co., New York, 1956.

Bakhvalov, N.S. [71] On the optimality of linear methods for operator approximation in convex classes of functions (in Russian), *Zh. Vychisl. Mat. Mat. Fiz.* **11** (1971), 1014–1018. [English transl.: *U.S.S.R. Computational Math. and Math. Phys.* **11** (1971), 244–249.]

Berlekamp, E.R. [68] *Algebraic Coding Theory,* McGraw-Hill, New York, 1968.

Bojańczyk, A. [80] Complexity of solving linear systems in different models of computation. [Submitted for publication.]

Borodin, A., and Munro, I. [75] *The Computational Complexity of Algebraic and Numeric Problems,* American Elsevier, New York, 1975.

De Millo, R.A., Dobkin, D., and Lipton, R.J. [78] Even data bases that lie can be compromised, *IEEE Trans. on Software Engineering* **SE-4** (1978), 73–75.

Denning, D.E., Denning, P.J., and Schwartz, M.D. [79] The tracker: a threat to statistical database security, *ACM Trans. on Database Systems* **4** (1979), 76–96.

Dobkin, D., Jones, A.K., and Lipton, R.J. [79] Secure databases: Protection against user influence, *ACM Trans. on Database Systems* **4** (1979), 97–106.

Gal, S., and Micchelli, A.C. [80] Optimal sequential and non-sequential procedures for evaluating a functional, *App. Anal.* **10** (1980), 105–120.

Garey, M.R., and Johnson, D.S. [79] *Computers and Intractability,* W.H. Freeman, San Francisco, California, 1979.

Gentleman, W.H. [78] Some complexity results for matrix multiplication on parallel processors, *J. Assoc. Comput. Mach.* **25** (1978), 112–115.

GTOA, *A General Theory of Optimal Algorithms,* see Traub and Woźniakowski [80a].

Hamming, R.W. [80] *Coding and Information Theory,* Prentice-Hall, Englewood Cliffs, New Jersey, 1980.

Johnson, D.S. [74] Fast algorithms for bin packing, *J. Comptr. Syst. Sci.* **8** (1974), 272–314.

Kiełbasiński, A. [81] Iterative refinement for linear systems in variable-precision arithmetic, *BIT* **21** (1981), 97–103.

Kornejčuk, N.P. [76] *External Problems of Approximation Theory,* Nauka, Moscow, 1976 (in Russian).

Lorentz, G.G. [66] *Approximation of Functions,* Holt, New York, 1966.

Marchuk, A.G., and Osipenko, K.U. [75] Best approximation of functions specified with an error at a finite number of points (in Russian), *Mat. Zameki* **17** (1975), 359–368. [English transl.: *Math. Notes* **17** (1975), 207–212.]

Marschak, J., and Radner, R. [72] *Economic Theory of Teams,* Cowles Foundation for Research in Economics, Monograph no. 22, Yale University Press, New Haven, 1972.

Micchelli, C.A. [75] *Optimal estimation of linear functionals,* IBM Research Rep. 5729 (1975).

Micchelli, C.A., and Miranker, W.L. [75] High order search methods for finding roots, *J. Assoc. Comput. Mach.* **22** (1979), 51–60.

Micchelli, C.A., and Rivlin, T.J. [77] A survey of optimal recovery, in *Optimal Estimation in Approximation Theory* (C.A. Micchelli and T.J. Rivlin, eds.), 1–54, Plenum Press, New York, 1977.

Miller, Webb [75] Computational complexity and numerical stability, *SIAM J. Comput.* **4** (1975), 97–107.

Pearl, J. [76] Theoretical bounds on the complexity of inexact computations, *IEEE Trans. Inform. Theory,* vol. **IT-22** (1976), 580–586.

Rabin, M.O. [72] Solving linear equations by means of scalar products, in *Complexity of Computer Computations* (R.E. Miller and J.W. Thatcher, eds.), 11–20, Plenum Press, New York, 1972.

Rivest, R.L., Meyer, A.R., Kleitman, D.J., Winklmann, K., and Spencer, J. [80] Coping with errors in binary search procedures, *J. Comptr. Syst. Sci.* **20** (1980), 396–404.

Rivest, R.L., and Vuillemin, J. [76] On recognizing graph properties from adjacency matrices, *Theoret. Comptr. Sci.* **3** (1976), 371–384.

Schönhage, A., and Strassen, V. [71] Schnelle Multiplikation grosser Zahlen, *Comput.* **7** (1971), 281–292.

Shannon, C.E. [48] A mathematical theory of communications, *Bell Syst. Tech. J.* **27** (1948), 379–423, 623–656.

Sikorski, K. [80] Optimal solution of nonlinear equations satisfying a Lipschitz condition. [Submitted for publication.]

[81] Bisection is optimal. [To appear in *Num. Math.*]

Sukharev, A.G. [76] Optimal search for the roots of a function satisfying a Lipschitz condition (in Russian), *Zh. Vychisl. Mat. Mat. Fiz.* **16** (1976), 20–30 [English transl.: *U.S.S.R. Computational Math. and Math. Phys.* **16** (1976), 17–26.]

Thompson, C.D. [80] *A Complexity Theory of VLSI,* Dissertation, Carnegie-Mellon Univ. (1980).

Traub, J.F., and Woźniakowski, H. [80a] *A General Theory of Optimal*

Algorithms, Academic Press, New York, 1980.

[80b] *On the Optimal Solution of Large Linear Systems,* Dept. of Computer Science Rep., Columbia Univ. (1980).

[80c] Convergence and complexity of interpolatory – Newton iteration in a Banach space, *Comp. and Maths. with Appls.* **6** (1980), 385–400.

[82] Complexity of linear programming, *O.R. Letters* **1** (1982), 59–62.

Traub, J.F., Woźniakowski, H., and Yemini, Y. [81] *Statistical Security of a Statistical Data Base,* Dept. of Computer Science Rep., Columbia Univ. (1981).

Trojan, J.M. [80] *Optimal Iterative Methods for the Solution of Nonlinear Equations,* Doctoral Thesis (in Polish), Institute of Informatics, University of Warsaw (1980).

Wasilkowski, G.W. [79] *Any Iteration for Polynomial Equations Using Linear Information has Infinite Complexity,* Dept. of Computer Science Rep., Carnegie-Mellon Univ. (1979). [To appear in *Th. Comp. Sci.*]

Werschulz, A.G. [82] Measuring uncertainty without a norm. [Submitted for publication.]

Yao, A.C. [79] Some complexity questions related to distributive computing. *Proc. 11th Ann. ACM Symp. on Theory of Computing,* Atlanta, Georgia (1979), 209–213.

[80] New algorithms for bin packing, *J. Assoc. Comput. Mach.* **27** (1980), 207–227.

AUTHOR INDEX

SUBJECT INDEX